Advanced Valuation

Macmillan Building and Surveying Series

Series Editor: IVOR H. SEELEY
  Emeritus Professor, The Nottingham Trent University

List continued overleaf

Advanced Valuation

DIANE BUTLER

M. Phil., F.R.I.C.S.
Senior Lecturer in Valuation
Department of Surveying
Nottingham Trent University

DAVID RICHMOND

F.R.I.C.S., F.S.V.A.
Principal Lecturer in Valuation
Department of Surveying
Nottingham Trent University

MACMILLAN

First published 1990 by
MACMILLAN PRESS LTD
Houndmills, Basingstoke, Hampshire RG21 6XS
and London
Companies and representatives
throughout the world

ISBN 0–333–47150–4

A catalogue record for this book is available
from the British Library.

11 10   9   8   7   6   5   4   3
03   02   01   00 99 98 97 96

Printed in Malaysia

Diane would like to dedicate this book to her mother and the memory of her father, and David to the memory of his parents

Contents

Contents

Preface

Following our individual texts, aimed at elementary and intermediate students of valuation, it seemed appropriate to produce a combined effort for advanced level students of the subject. The students, with whom we are in daily contact, expressed the need for such a book.

The format follows its predecessors, giving consideration to typical valuation examination questions. Some of the suggested answers are more detailed than would be expected of an examination candidate, but the opportunity has been taken to explain in greater detail particular points that, in our experience, students have had difficulty in comprehending. Where answers are extended in this way, it is usually noted in the text, so that students may judge what is expected of them.

The text is arranged in topic areas, with questions and suggested answers shown separately. The phrase 'suggested answers' is used deliberately, since, in valuation, there is rarely a precise answer. Valuers may calculate valuation figures quite precisely, but, in practice, this is usually to establish their negotiating position, or to ascertain the parameters within which they are able to agree a figure. Similarly, in an examination situation, an examiner would be most surprised if the majority of candidates produced the same answer. The examiner is seeking an approach based upon sound valuation principles, rather than an exact mathematical answer.

It is suggested that this book may be used as a general guide in the areas of valuation that are considered, and/or as a revision aid. If used as the latter, the most benefit will probably be derived if individual questions are attempted initially without reference to the answer, before checking with this that the approach adopted is correct.

Unlike our previous books, greater consideration is given in this instance to questions which require discursive or essay-type answers.

We hope all students of valuation derive some help from this text and we wish them well.

We would like to thank the following people: Professor Ivor Seeley, the Series Editor, for once again advising and encouraging us; John Bailey and Steve Tyler for the use of some questions; and Mark Powell for information from his Nottingham Polytechnic dissertation.

We are particularly grateful to Mrs Susan Williams for her careful production of the manuscript, including the correction of our errors.

It has been a pleasant experience to embark upon a joint venture with a colleague and when that colleague is also a friend, it is all the more rewarding.

Nottingham   Diane Butler
Autumn 1989   David Richmond

1 Analysis and General Valuations

While the majority of chapters in this book will deal with valuations of a complex nature often involving legislation, it is necessary to appreciate the basic technique of analysing information and its application.

The purpose of analysing recent transactions is to obtain either current rental values of certain property types or initial yields (or, in some cases, both).

The questions used in this chapter which require the analysis of a recent transaction to obtain current rental value make use of the following techniques:

(i) The calculation of the annual equivalent of a premium to add to a passing rent, giving rental value. It is recommended that the annual equivalent is obtained from the viewpoint of both parties to the transaction, and averaged.

(ii) The carrying out of a surrender and renewal of lease calculation, where a lessee wishes to surrender a short unexpired term and increase his security with a longer period of time. Again, calculations from both viewpoints (landlord and tenant) should be undertaken and the figure for rental value averaged. The principle adopted is that, in both cases, the market value of the current situation should equate with the market value of the proposed situation.

Some of the valuations in this chapter are affected by landlord and tenant legislation, notably the *Landlord and Tenant Acts 1927 and 1954* and the *Law of Property Act 1969*. It may be necessary to calculate compensation for tenant's improvements, compensation for the loss of security of tenure and to take account of the effect of tenant's improvements upon subsequent rent calculations.

The reader is recommended to refer to chapter 3 of *Applied Valuation* by Diane Butler (Macmillan, 1987) before tackling these questions.

ANALYSIS AND GENERAL VALUATIONS – QUESTIONS

1.1. Smith is the lessee of a single storey shop having frontage to a shopping street of 5 m and a depth of 24 m. The current 14 year lease on internal repairing terms at a fixed rent of £1750 per annum has 4 years unexpired. When this lease was granted, Smith surrendered a full repairing and insuring lease at a rent of £1250 per annum with 4 years unexpired.

Eight years ago, Smith carried out improvements with the landlord's consent at a cost of £5000, the effect of these being to have increased the rental value by 12 per cent both then and today. It is considered that rental values of shops in this area have increased during the past 10 years by 15 per cent per annum.

Smith now wishes to extend the shop at the rear by a depth of 7 metres which would cost £12 000; this will have the same rental value per m^2 as the rear zone of the existing accommodation. The freeholder would consent to this provided that Smith surrenders his present lease and takes a 20 year full repairing and insuring lease at a rent to be agreed but with rent review clauses to full rental value after 10 and 15 years. The proposed improvement would be carried out by Smith as a term of the new lease. The freeholder has, however, agreed to pay Smith suitable compensation for the improvement undertaken 8 years ago.

Considering the present position from Smith's viewpoint only, advise him on a reasonable rent to be paid per annum for the proposed new lease.

1.2. White Components Limited occupies factory A, which it holds on a full repairing and insuring lease with 3 years unexpired at a rent of £6500 per annum. The company wishes to surrender its existing lease and take a new lease on similar terms for a period of 20 years. The freeholder is agreeable to this provided that the rent is reviewed every 5 years and adjusted to the full rental value applicable at the time of review. Factory A is one of four single-storey factories with office blocks attached.

You have obtained the following details about these premises in relation to floor areas and transactions.

Factory	Net floor space (square metres)		Transaction details
	Factory	Offices	
A	700	200	
B	600	125	Let for 5 years, 2 years ago on full repairing and insuring terms at £5000 per annum The lease was sold recently for £2825
C	750	150	Freehold sold with vacant possession last month for £104 500
D	700	250	Let last month for 5 years at a rent of £10 500 per annum on internal repairing terms, subject to the payment of £3700 on entry

It is considered that rental values of factories per m^2 are one-third of the values of offices per m^2.

Analyse the above transactions and, acting between the parties, calculate the rent per annum to be reserved for the first 5 years under the new lease of factory A.

1.3. Your client has obtained planning permission to redevelop a site and this will necessitate acquiring all the interests in the following existing freehold properties on the site.

(a) A shop, frontage 10 m and depth 16 m, with one storey offices over (same dimensions as shop). The shop was let 2 years ago on a 7 year internal repairing lease at a rent of £7000 per annum.

A similar shop, frontage 15 m and depth 12 m, was recently let on a 7 year full repairing and insuring lease at a rent of £9750 per annum, subject to the payment of a premium of £10 000 on entry.

The offices were let 2 years ago on a 7 year internal repairing lease at a rent of £6500 per annum; half of the area, however, was sublet last month on a 3 year internal repairing lease at a rent of £3500 per annum; subject to the payment of a premium of £1000 on entry. The current rateable value of the whole is £12 000.

(b) A single-storey workshop let to the lessee 2 years ago on a 12 year full repairing and insuring lease at a rent of £4000 per annum, subject to 6 year rent reviews to current market value at the time of review. The lessee improved the property with the freeholder's consent last year at a cost of £5000, which has contributed £750 per annum to the current net rental value of £6000 per annum.

The current rateable value of the premises is £3750.

You are required to prepare valuations (assuming all parties are prepared to sell) indicating the amounts which your clients should be prepared to pay to all owners and occupiers to obtain vacant possession of the properties.

1.4. Your clients are the directors of a professional football club and they wish to provide an indoor training centre and social club near the ground; this will necessitate acquiring all the interests in the following existing properties on the site:

(a) Industrial premises, 350 m^2 net, let by the freeholder to the lessee 2 years ago on a 7 year internal repairing lease at a rent of £6000 per annum.

The current rateable value is £4000.

Similar premises, 400 m^2 net, have been let recently on a 7 year full repairing and insuring lease at a rent of £8200 per annum. The lessee surrendered a 7 year internal repairing lease, having 2 years unexpired at a rent of £6000 per annum.

(b) Shop premises, frontage 12 m and depth 20 m, with first floor offices of 240 m^2 net over. The shop was let by the freeholder to the lessee on a 7 year internal repairing lease 3 years ago at a rent of £18 000 per annum, and the offices are occupied by the freeholder.

A similar shop, frontage 10 m and depth 18 m, has been let recently on a 7 year full repairing and insuring lease at a rent of £16 000 per annum, subject to the payment of a premium of £5000 on entry.

Upper floors currently let at £35 per m² on full repairing and insuring terms.

The current rateable value of the combined premises is £20 000.

You are required to prepare valuations (assuming all parties are prepared to sell) indicating the amount which your clients should be prepared to pay to all owners and occupiers to obtain vacant possession of the properties.

1.5. Happydays Toys Limited occupies shop A, which it holds on a full repairing and insuring lease with 2 years unexpired at a rent of £10 000 per annum. The company wishes to surrender its existing lease and take a new one for 10 years with a rent review to current market rent after 5 years. The freeholder is agreeable, provided the company improves the property at the commencement of the new lease. This will cost £20 000 and will increase the rental value by £2000 per annum.

Shop A is one of three ground floor shops with one storey offices over, and you have obtained the following details about these premises in relation to floor areas and transactions.

Shop	Frontage and depth (metres)	Offices (net floor space — m²)	Transaction details
A	8 frontage 16 depth	100	Details given above
B	9 frontage 18 depth	150	Let last month for 5 years on full repairing and insuring terms at a rent of £16 500 per annum, subject to the payment of £10 000 on entry
C	10 frontage 15 depth	120	Freehold sold last month for £308 750

It is considered that rental values per m² of offices are 30 per cent of Zone A rents (5 m depth) per m².

Analyse the above transactions, and, acting between the parties, calculate the rent per annum to be reserved for the first 5 years under the new lease of shop A.

ANALYSIS AND GENERAL VALUATIONS – SUGGESTED ANSWERS

Question 1.1

The first step in answering this question is to analyse the surrender and renewal of lease transaction, which took place 10 years ago, to establish the full rental value per annum at that time. This should be carried out from both freehold and lease-hold viewpoints with a present equals proposed interest type calculation. (Refer to chapter 3 of *Applied Valuation* by Diane Butler (Macmillan, 1987).)

Let rental value on full repairing and insuring terms = £X.

Landlord's present interest

Rent received	£1 250 pa	
YP 4 years at $6\frac{1}{2}$ per cent [see note 1]	3.426	£4 282
Reversion to full net rental value	£X	
YP in perpetuity deferred 4 years at 7 per cent [see note 1]	10.899	10.899X
Capital value		£4 282 + 10.899X

Note

1: Traditional yield pattern. Initial (all risks) yield for this type of property assumed to be 7 per cent. The 4 year term considered to be more secure than reversion so yield is $\frac{1}{2}$ per cent less.

Landlord's proposed interest

Rent received	£1 750 pa	
Less External repairs and insurance [see note 1]	0.075X	
Net income	£1 750 − 0.075X pa	
YP 14 years at 12 per cent [see note 2]	6.628	£11 599 − 0.4971X
Reversion to full net rental value	£X	
YP in perpetuity deferred 14 years at 7 per cent	5.54	5.54X
Capital value		£11 599 + 5.0429X

Notes

1: It is assumed that the allowance for external repairs and insurance would be $7\frac{1}{2}$ per cent of full net rental value, i.e. $0.075X$.

2: Because the fixed income is for a long (14 year) period, it is inflation-prone and a high yield is used to reflect this.

$$\begin{aligned} \text{Present} &= \text{Proposed} \\ \pounds 4\,282 + 10.899X &= \pounds 11\,599 + 5.0429X \\ X &= \pounds 1\,250 \text{ pa} \end{aligned}$$

Lessee's present interest

Full net rental value	$\pounds X$
Less rent paid	$\pounds 1\,250$ pa
Profit rent	$\pounds X - 1\,250$ pa
YP 4 years at 8 per cent	
and 3 per cent (tax 40 per cent)	2.09
Capital value	$\pounds 2.09X - 2\,612$

Lessee's proposed interest

Full rental value on internal repairing terms	
[see note 1]	$\pounds 1.075X$
Less rent paid	$\pounds 1\,750$ pa
Profit rent	$\pounds 1.075X - 1\,750$ pa
YP 14 years at 8 per cent	
and 3 per cent (tax 40 per cent)	5.632
Capital value	$\pounds 6.0544X - 9\,856$

Note

1: The full rental value on internal repairing terms is the net rental value plus external repairs and insurance, i.e. $1.075X$.

$$\begin{aligned} \text{Present} &= \text{Proposed} \\ \pounds 2.09X - 2\,612 &= \pounds 6.0544X - 9\,856 \\ X &= \pounds 1\,827 \text{ pa} \end{aligned}$$

Averaging between £1250 and £1827 gives a net rental value of say £1550 pa 10 years ago. This figure needs to be updated by 10 years to give a current full rental value.

i.e. Rental value 10 years ago   £1 550 pa
      Amount of £1 2 years at
          15 per cent   1.3225

      Rental value 8 years ago   £2 050 pa
      Value of improvement
          12 per cent of £2050   £   246 pa

                                 £2 296 pa

      Amount of £1 8 years at
          15 per cent   3.059

                                 £7 023

      Current net rental value say   £7 000 pa

The £7000 represents the value of the unimproved property, say y, plus the value of the 8 year old improvement $0.12y$.

So that the value of the unimproved shop is

$$\frac{7\,000}{1.12} = £6\,250$$

Thus the 8 year old improvement is currently worth £750 pa. As part of the agreement for the granting of the new lease, the landlord has agreed to pay Smith suitable compensation for the 8 year old improvement. This is considered to be his entitlement under the *Landlord and Tenant Act 1927* — the lesser of:

(i) The present reasonable cost of carrying out the improvements less any deduction necessary for the cost of putting the improvement into reasonable repair. This is subject to the liabilities of the lease; in this case, internal repairing.

    i.e. Cost 8 years ago   £5 000
        Amount of £1 − 8 years at
        12 per cent [see note 1]   2.476

                                 £12 380
       Less making good repair say   £1 000

                                 £11 380

Note

1: Assumed that building costs increased at an average rate of 12 per cent over the last 8 years

and

(ii) The net increase in value of the landlord's interest attributable to the improvement.

i.e.   Rental value of improvement   £750 pa
       YP in perpetuity at 7 per cent   14.28

say   £10 700

The landlord will pay to Smith compensation of £10 700.

The next stage of the calculation is to zone the current rental value of £7000 pa to find the remainder zone value to apply to the new extension.

Using a 'halving back' method and two 5 m zones and a remainder

$$
\begin{aligned}
&\text{Let Zone A rental value per m}^2 = \pounds X \\
&\text{Zone A} \quad = 5 \times 5 \times X \quad\quad\quad\ = 25X \\
&\text{Zone B} \quad = 5 \times 5 \times \tfrac{1}{2}X \quad\quad = 12.5X \\
&\text{Remainder} = 5 \times 14 \times \tfrac{1}{4}X \quad = 17.5X \\
&\hspace{6.5cm} \overline{55X}
\end{aligned}
$$

$$
\text{So that } X = \frac{7\,000}{55} = \pounds127 \text{ per m}^2
$$

The rear zone is worth $\tfrac{1}{4}$ of £127 = £32 per m^2 so that the extension has a net rental value of 5 m x 7 m x £32 = £1 120 per annum.

Information has now been obtained to carry out the last step of the calculation — Smith's present and proposed interests.

Let £X = rent to be paid per annum for the first 10 years.

Smith's present interest

Rental value on internal repairing terms [see note 1]	£7 525 pa	
Less rent paid	£1 750 pa	
Profit rent	£5 775 pa	
YP 4 years 8 per cent and 3 per cent (tax 40 per cent)	2.09	£12 070
Rental value of 8 year old improvement [see note 2]	£750 pa	
YP 10 years 8 per cent and 3 per cent (tax 40 per cent) [see note 2]	4.437	
x PV of £1 in 4 years at 8 per cent	0.735	£2 446
Capital value		£14 516

Notes

1: The rental value on internal repairing terms is the net rental value and the cost of external repairs and insurance (say $7\frac{1}{2}$ per cent of rental value, i.e. $7\frac{1}{2}$ per cent of £7000 = £525 pa). Thus, the figure is £7000 + £525 = £7525 pa.

2: It is assumed that the tenant may be granted a new lease after the expiration of the present one. The value of the tenant's improvements carried out during the previous 21 years should be disregarded on the granting of the new lease, in accordance with the *Law of Property Act 1969*. The tenant, therefore, has an immediate profit rent of £750 pa. It is, further, assumed that this provision applies to rent reviews so that with 5 year rent reviews, Smith would have the benefit of reduced rent for a further 10 years. (Note: If the lease was silent about the effect of improvements at rent review, the first rent review would take account of the value of the improvement. Refer to *Ponsford v. HMS Aerosols Ltd, 1978*.)

Smith's proposed interest

Improved net rental value [see note 1]	£8 120 pa
Less rent paid	£X pa
Profit rent	£8 120 − X pa
YP 10 years at 8 per cent and 3 per cent (tax 40 per cent)	4.437
	£36 028 − 4.437X
Plus Compensation for 8 year old improvement (see earlier)	£10 700
	£46 728 − 4.437X
Less Cost of extension	£12 000
Capital value	£34 728 − 4.437X

Note

1: The net rental value is £7000 pa and the extension is valued at £1 120 pa. Thus, total is £8120 pa.

$$\text{Present} = \text{Proposed}$$
$$£14\,516 = £34\,728 - 4.437X$$
$$X = £4\,555 \text{ pa}$$

Rent for the first ten years, say £4500 pa.

Question 1.2

The answer to this question entails initially analysing transactions relating to factories B, C and D to obtain rental value per m² of factory and office accommodation and also initial yields for freehold and leasehold interests. The order of analysing transactions may vary, but it is suggested that the following may be undertaken:

1: Analyse transaction of factory D to obtain a rent per m² for offices and factory.
2: Apply these rentals to the freehold transaction of factory C to obtain an initial yield.
3: Analyse transaction of factory B to check leasehold yield.
4: Apply results to factory A.

The shortcoming of this type of situation is that it is necessary to assume freehold and leasehold yields to calculate the annual equivalent of the premium in the transaction of factory D.

If this is subsequently inconsistent with the yields calculated for factories B and C, the process may have to be repeated.

Analysis of factory D

Rent passing   £10 500 pa
Plus Annual equivalent of premium [see note 1]

$$\text{Freeholder} \quad \frac{£3\,700}{\text{YP 5 years at 8 per cent}}$$

$$= \frac{£3\,700}{3.993} = £926$$

$$\text{Lessee} \quad \frac{£3\,700}{\substack{\text{YP 5 years at 9 per cent} \\ \text{and 3 per cent} \\ \text{(tax 40 per cent)}}}$$

$$= \frac{£3\,700}{2.476} = £1\,494$$

Average   = £1 210 pa

Rental value on internal repairing terms say £11 700 pa

Note

1: The annual equivalent of the premium of £3700 has been calculated from both freehold and leasehold viewpoints, averaged and added to rent passing to give the rental value of £11 700 pa. An initial yield of 8 per cent has been assumed for the freehold interest and 9 per cent and 3 per cent (tax 40 per cent) for the lease-hold interest.

The rental value on internal repairing terms needs to be split between factory and office accommodation, factory space having $\frac{1}{3}$ of the value per m^2 of office space.

$$\text{Let rental value per m}^2 \text{ of offices} = £X$$
$$\text{Then } 250X + (700 \times 0.3334X) = £11\,700$$

$$X = \frac{11\,700}{483.4} = £24.2 \text{ per m}^2$$

So that the rental value of offices is £24 per m^2 and factory space £8 per m^2 on internal repairing terms.

Factory C

Applying the rents per m^2 calculated above:

Factory 750 m^2 × £8	= £6 000
Offices 150 m^2 × £24	= £3 600

Rental value on internal repairing terms = £9 600 pa.

Assume that external repairs and insurance cost 15 per cent of net rental value per annum, then net rental value

$$= \frac{9\,600}{1.15} = £8\,350 \text{ pa}$$

$$\text{Initial yield} = \frac{\text{Net rental value}}{\text{Capital value}}$$

$$= \frac{8\,350}{104\,500} = 0.0799$$

say 8 per cent

A freehold initial yield of 8 per cent is consistent with that used in the analysis of factory D.

Analysis of factory B

Applying the rents per m² calculated above:

Factory 600 m² × £8	= £4 800
Offices 125 m² × £24	= £3 000
Rental value on internal repairing terms	£7 800 pa
Less external repairs and insurance	£1 020 pa
[see note 1]	
Rental value on full repairing and insuring terms	£6 780 pa
Less Rent paid	£5 000 pa
Profit rent	£1 780 pa

Note

1: The cost of external repairs and insurance is assumed to be 15 per cent of net rental value so that net rental value is

$$\frac{7\,800}{1.15} = £6\,780 \text{ pa}$$

$$\text{Now YP} = \frac{\text{Capital value}}{\text{Profit rent}} = \frac{2\,825}{1\,780} = 1.587$$

This Years' Purchase of 1.587 is for 3 years at *i* (the unknown initial yield), sinking fund at 3 per cent net and tax at 40 per cent.

Using Years' Purchase Dual Rate tables, 3 per cent net (tax 40 per cent) for 3 years, the nearest YP figure to 1.587 is 1.5893, given at 9 per cent.

So that initial leasehold yield is 9 per cent, which is consistent with earlier calculations.

Note: If the above 'short-cut' method was not used, '*i*' would be obtained from

$$\text{YP} = \frac{1}{i + s\left(\dfrac{1}{1 - t}\right)}$$

$$1.587 = \frac{1}{i + \left(\dfrac{0.03}{1.03^3 - 1} \times \dfrac{1}{0.6}\right)}$$

$$1.587 = \frac{1}{i + 0.539}$$

$$\text{So } i = \frac{1 - (1.587 \times 0.539)}{1.587} = 0.09 = 9 \text{ per cent}$$

Applying results to factory A

$$\begin{array}{ll}
\text{Factory 700 m}^2 \times \text{ £8} & = \text{£5 600} \\
\text{Offices 200 m}^2 \times \text{£24} & = \text{£4 800}
\end{array}$$

Rental value on internal repairing terms = £10 400

$$\text{Net rental value } = \frac{\text{£10 400}}{1.15} = \text{say £9 040 pa}$$
[see note 1]

Note

1: An allowance has been made as before for external repairs and insurance at 15 per cent of net rental value pa.

The final step in answering this question is to obtain the annual rent for the first 5 years of the new lease by carrying out a surrender and renewal of lease calculation. This is from both freehold and leasehold viewpoints, using a 'present interest = proposed interest' approach.

Let rental to be paid on full repairing and insuring terms = £X pa.

Landlord's present interest

Rent received	£6 500 pa	
YP for 3 years at 7 per cent [see note 1]	2.624	£17 056
Reversion to full net rental value	£9 040 pa	
YP in perpetuity deferred 3 years at 8 per cent [see note 1]	9.923	£89 703
		£106 759
Capital value say		£106 800

Note

1: 8 per cent has been established as the initial yield for freehold interests from the previous analyses and is used for the reversion. The term is considered to be more secure so a lower yield of 7 per cent is adopted.

Landlord's proposed interest

Rent received	£X	
YP for 5 years at 7 per cent	4.1	4.1X
Reversion to full net rental value	£9 040 pa	
YP in perpetuity deferred 5 years at 8 per cent	8.507	£76 903
Capital value		4.1X + 76 903

$$\begin{aligned}
\text{Present} &= \text{Proposed} \\
£106\,800 &= 4.1X + £76\,903 \\
X &= £7\,291 \\
\text{say} &= £7\,300 \text{ pa}
\end{aligned}$$

Lessee's present interest

Full net rental value	£9 040 pa
Less rent paid	£6 500 pa
Profit rent	£2 540 pa
YP 3 years at 9 per cent and 3 per cent (tax 40 per cent)	1.589
Capital value	£4 036
say	£4 000

Lessee's proposed interest

Full net rental value	£9 040 pa
Less rent paid	£X pa
Profit rent	£9 040 − X pa
YP for 5 years at 9 per cent and 3 per cent (tax 40 per cent)	2.476
Capital value	£22 383 − 2.476X

$$\begin{aligned}
\text{Present interest} &= \text{Proposed interest} \\
£4\,000 &= £22\,383 - 2.476X \\
X &= £7\,425 \text{ pa}
\end{aligned}$$

The recommended rent pa may be averaged between £7300 and £7425 — say, £7360 pa.

Question 1.3

(a) The first step to be taken in this question is to analyse the recent rental transaction of the similar shop to obtain its current net rental value.

The annual equivalent of the premium should be obtained from both landlord
and tenant viewpoints and the average added to the rent passing.

Analysis of similar shop

Rent passing	£9 750 pa
Plus Annual equivalent of premium	

$$\text{Freeholder} = \frac{£10\,000}{\text{YP 7 years at 6 per cent}}$$
$$[\text{see note 1}]$$

$$= \frac{£10\,000}{5.582} = £1\,791$$

$$\text{Lessee} = \frac{£10\,000}{\substack{\text{YP 7 years at 7 per cent} \\ \text{and 3 per cent} \\ \text{(tax 40 per cent)}}}$$

$$= \frac{£10\,000}{3.478} = £2\,875$$

	Average	= £2 333 pa
Full net rental value		= £12 083 pa
	say	£12 000 pa

Note

1: 6 per cent is considered to be a realistic initial yield for the freehold of a shop.

The net rental value should be zoned using two 5 metre zones and a remainder and
a 'halving back' principle to obtain a Zone A rent per m^2.

Let Zone A rent per m^2 = £X

$15 \times 5 \times X$	$= 75X$
$15 \times 5 \times \frac{1}{2}X$	$= 37.5X$
$15 \times 2 \times \frac{1}{4}X$	$= \underline{7.5X}$
	$120X$

$$X = £\frac{12\,000}{120} = £100 \text{ per } m^2$$

So that Zone A rent per m^2 = £100

Applying this to the subject shop:

$$10 \times 5 \times £100 = £5\,000$$
$$10 \times 5 \times £\ \ 50 = £2\,500$$
$$10 \times 6 \times £\ \ 25 = £1\,500$$

Net rental value	= £9 000 pa

Valuation

Freeholder

Rent received	£7 000 pa	
Less external repairs and insurance [see note 1]	£ 900 pa	
Net income	£6 100 pa	
YP 5 years at 5 per cent [see note 2]	4.329	£26 407
Reversion to full net rental value	£9 000 pa	
YP in perpetuity deferred 5 years at 6 per cent	12.454	£112 086
		£138 493
Capital value say		£138 500

Notes

1: The allowance for external repairs and insurance per annum is 10 per cent of net rental value.
2: A traditional valuation method has been used. The income during the lease is considered to be secure so that a yield 1 per cent less than the initial yield is used.

Lessee

Full rental value on internal repairing terms [see note 1]	£9 900 pa
Less rent paid	£7 000 pa
Profit rent	£2 900 pa
YP 5 years at 7 per cent and 3 per cent (tax 40 per cent)	2.605
	£7 554
Capital value say	£7 550

Note

1: The full rental value on internal repairing terms is £9 000 + £900 = £9 900 per annum.

The valuation of the three interests in the office — freeholder, lessee and sublessee — should now be undertaken.

It is necessary to analyse last month's subletting to obtain the full rental value per m^2 for offices. This is achieved by calculating the annual equivalent of the premium from both freeholder and lessee viewpoints; taking the average and adding it to the rent passing.

Rent passing   £3 500 pa
Plus Annual equivalent of premium

$$\text{Freeholder} \quad £ \; \frac{1\,000}{\text{YP 3 years at 7 per cent}}$$
[see note 1]

$$= \frac{£1\,000}{2.624} = £381$$

$$\text{Lessee} \quad £ \; \frac{1\,000}{\substack{\text{YP 3 years at 8 per cent} \\ \text{and 3 per cent} \\ \text{(tax 40 per cent)}}}$$

$$= \frac{£1\,000}{1.615} = £619$$

Average   = £500 pa

Full rental value on internal repairing terms   = £4 000 pa
Rent per m^2 on internal repairing terms

$$= \frac{4\,000}{80} \quad \text{[see note 2]} \quad = £50 \text{ per } m^2$$

Notes

1: It is considered that 7 per cent is an appropriate initial yield for the office.

2: The sublet portion has an area of 50 per cent of 10 m × 16 m = 80 m^2.

The total rental value on internal repairing terms is 10 m × 16 m × £50 per m^2 = £8 000 per annum.

The rental value on full repairing and insuring terms, assuming an allowance of 10 per cent of net value for external repairs and insurance, is 8 000/1.10 = £7 270 per annum.

Valuation of office

Freeholder

Rent received	£6 500 pa	
Less external repairs and insurance [see note 1]	£ 730 pa	
Net income pa	£5 770 pa	
YP 5 years at 6 per cent [see note 2]	4.212	£24 303
Reversion to full net rental value	£7 270 pa	
YP in perpetuity deferred 5 years at 7 per cent	10.186	£74 052
		£98 355
Capital value say		£98 400

Notes

1: The allowance for external repairs and insurance is 10 per cent of the net rental value of £7 270 per annum, say £730 per annum.

2: The income during the lease is considered to be secure so that a yield 1 per cent less than the initial yield is used.

Lessee

Rent received from sublessee	£3 500 pa	
Full rental value on internal repairing terms of occupied area [see note 1]	£4 000 pa	
	£7 500 pa	
Less Rent paid	£6 500 pa	
Profit rent	£1 000 pa	
YP 3 years at 8 per cent and 3 per cent (tax 40 per cent)	1.615	£1 615
Reversion to full rental value on internal repairing terms [see note 2]	£8 000 pa	
Less rent paid	£6 500 pa	
Profit rent	£1 500 pa	
YP 2 years at 8 per cent and 3 per cent (tax 40 per cent)	1.11	
x PV of £1 in 3 years at 8 per cent	0.794	£1 322
		£2 937
Capital value say		£2 950

Notes

1: The occupied area for the first 3 years is 50 per cent of 10 m × 16 m = 80 m^2 worth £50 per m^2 = £4000 per annum.

2: For the last 2 years of the lease, the lessee has possession of the whole area and could occupy himself or theoretically let it on internal repairing terms. In either situation it is worth £8 000 per annum.

Sublessee

The value of the sublease is equivalent to the premium of £1 000 paid last month. This is because the premium is equal to 3 years' capitalised profit rent.

The occupants of the shop and offices may seek from the developer compensation for their loss of occupancy and this may be calculated in accordance with the provisions of the *Landlord and Tenant Act 1954*. The leases and sublease are less than 14 years duration so that compensation will be based on 3 × Rateable value.

The total rateable value is £12 000, and it is suggested that this is apportioned between each occupant on the basis of:

$$\frac{\text{Full rental value of occupied portion}}{\text{Full rental value of the whole}}$$

Full net rental value is £9 000 (shop) + £7 270 (office) = £16 270 per annum.

So that compensation may be:

Lessee of shop $\dfrac{9\,000}{16\,270}$ × 3 × £12 000   = £19 914

Lessee of office $\dfrac{3\,635}{16\,270}$ × 3 × £12 000   = £8 043

Sublessee of office $\dfrac{3\,635}{16\,270}$ × 3 × £12 000 = £8 043

                        Total   £36 000

Because of the complex calculation, it may be advisable to conclude part (a) with a summary of payments for shop and office:

i.e. Freeholder	Shop	£138 500	
	Office	£98 400	£236 900
Lessee of shop		£7 550	
	Compensation say	£19 900	£27 450
Lessee of office		£2 950	
	Compensation say	£8 000	£10 950
Sublessee of office		£1 000	
	Compensation say	£8 000	£9 000
	Total		£284 300

(b) When valuing the freehold and leasehold interests in this question, account must be taken of the improvement carried out by the tenant and its effect on subsequent rent. It is assumed that the lease provides for disregarding the tenant's improvements when reviewing rent. If this is the case the review in 4 years' time will disregard the improvement. Because of the '21 year rule', if a new lease is created in 10 years' time with further 6 year reviews (unusual!), the tenant may have 'disregard' benefits for two further periods, i.e. 12 years.

(If the lease was silent regarding how tenant's improvements should be treated on review, *Ponsford v. HMS Aerosols Ltd 1978* would apply, and on reviews the improvement would be taken into account.)

Freeholder

Rent passing	£4 000 pa	
YP 4 years at 7 per cent	3.387	£13 548
Reversion to unimproved net rental value [see note 1]	£5 250 pa	
YP 18 years at 8 per cent [see note 1]	9.372	
× PV of £1 in 4 years at 8 per cent	0.735	£36 164
Reversion to full rental value	£6 000 pa	
YP in perpetuity deferred 22 years at 8 per cent [see note 2]	2.299	£13 794
		£63 506
Capital value say		£63 500

Notes

1: The lessee may have the benefit of paying the unimproved rental value for 18 years dependent upon the circumstances explained earlier.

2: It is considered that 8 per cent is an appropriate freehold yield for a workshop.

Lessee

Full net rental value	£6 000 pa	
Less Rent paid	£4 000 pa	
Profit rent	£2 000 pa	
YP 4 years at 9 per cent and		
3 per cent (tax 40 per cent)	2.048	£4 096
Full net rental value	£6 000 pa	
Less Unimproved net rental value	£5 250 pa	
Profit rent	£ 750 pa	
YP 18 years at 9 per cent and		
3 per cent (tax 40 per cent)	6.204	
× PV of £1 in 4 years at 9 per cent	0.708	£3 294
		£7 390
Capital value say		£7 400

Plus Compensation under *Landlord and
Tenant Act 1954* [see note 1]

3 × Rateable value = 3 × £3750 = £11 250

£18 650

Note

1: It is considered that the tenant would seek compensation for loss of occupancy as provided in the *Landlord and Tenant Act 1954*. As the lease is less than 14 years, this will be 3 × Rateable value.

Summary

Freeholder	£63 500
Lessee	£18 650

Note

All parties involved in the shop, office and workshop might negotiate for amounts in excess of those calculated. They would hope to receive some proportion of the marriage gain realised when the development site is assembled. The question gives no information regarding development value, so marriage gain is incalculable.

Question 1.4

(a) The first step to be undertaken in this calculation is to obtain the current rental value per m^2 for industrial premises.

This will necessitate the surrender and renewal of lease transaction being analysed, using the present interest equals proposed interest principle previously described.

Let rental value on full repairing and insuring terms = £X.

Landlord

Present interest

Rent passing	£6 000 pa	
Less External repairs and insurance [see note 1]	0.1X	
Net income	£6 000 − 0.1X pa	
YP 2 years at 7 per cent [see note 2]	1.808	£10 848 − 0.1808X
Reversion to full net rental value	£X	
YP in perpetuity deferred 2 years at 8 per cent [see note 2]	10.717	£10.717X
Capital value		£10 848 + 10.536X

Notes

1: An allowance for external repairs and insurance has been made per annum at 10 per cent of net rental value.

2: An initial yield of 8 per cent is considered to be appropriate for the freehold interest in industrial premises. The income receivable during the 2 year term is considered to be secure and the yield is 1 per cent lower than the reversion.

Proposed interest

Rent passing	£8 200 pa	
YP 7 years at 7 per cent	5.389	£44 190
Reversion to full net rental value	£X	
YP in perpetuity deferred 7 years at 8 per cent	7.294	7.294X
Capital value		£44 190 + 7.294X

$$
\begin{aligned}
\text{Present} &= \text{Proposed} \\
£10\,848 + 10.536X &= £44\,190 + 7.294X \\
X &= £10\,284 \text{ pa}
\end{aligned}
$$

Lessee

Present interest

Full rental value on internal repairing terms [see note 1]	£1.1X
Less Rent paid	£6 000 pa
Profit rent	£1.1X − 6 000
YP 2 years at 9 per cent and 3 per cent (tax 40 per cent)	1.098
Capital value	£1.208X − 6 588

Note

1: The full rental value on internal repairing terms per annum is net rental value + external repairs and insurance, i.e. $X + 0.1X = 1.1X$.

Proposed interest

Full net rental value	£X
Less Rent paid	£8 200 pa
Profit rent	£X − 8 200 pa
YP 7 years at 9 per cent and 3 per cent (tax 40 per cent)	3.252
Capital value	£3.252X − 26 666

Present interest = Proposed interest
£1.208X − 6 588 = £3.252X − 26 666
X = £9 822 pa

Averaging between £10 284 and £9 822, the net rental value is considered to be £10 050, say £10 000 per annum.

The net rental value for industrial premises per m^2

$$= \frac{£10\,000}{400} = £25$$

Applying to the subject premises, the net rental value is 350 m^2 × £25 = £8 750 per annum.

The freehold and leasehold interests in the subject premises may now be valued.

Freeholder

Rent passing	£6 000 pa	
Less External repairs and insurance		
[see note 1]	£875 pa	
Profit rent	£5 125 pa	
YP 5 years at 7 per cent	4.1	£21 012
Reversion to full net rental value	£8 750 pa	
YP in perpetuity deferred 5 years		
at 8 per cent	8.507	£74 436
		£95 448
Capital value say		£95 500

Note

1: The allowance for external repairs and insurance per annum is 10 per cent of net rental value.

Lessee

Full rental value on internal repairing terms		
[see note 1]	£9 625 pa	
Less Rent paid	£6 000 pa	
Profit rent	£3 625 pa	
YP 5 years at 9 per cent and 3 per cent		
(tax 40 per cent)	2.476	
	£8 975	
Plus 3 × Rateable value [see note 2]	£12 000	
	£20 975	
Payment to lessee say	£21 000	

Notes

1: The rental value on internal repairing terms is the net rental value + external repairs and insurance = £8 750 + £875 = £9 625 per annum.

2: The lessee would expect to receive his entitlement to compensation under *Landlord and Tenant Act 1954* for denial of his occupancy. As the lease is for less than 14 years, his entitlement would be 3 × Rateable value.

(b) The first step is to calculate the net rental value per annum of the similar shop. This is obtained by calculating the annual equivalent of the premium from both freehold and leasehold viewpoints, taking the average and adding it to the rent passing, i.e.

Rent passing £16 000 pa
Plus Annual equivalent of premium

$$\text{Freeholder} \quad \frac{£5\,000}{\text{YP 7 years at 6 per cent}}$$
[see note 1]

$$= \frac{£5\,000}{5.582} = £896$$

$$\text{Lessee} \quad \frac{£5\,000}{\text{YP 7 years at 7 per cent and 3 per cent (tax 40 per cent)}}$$

$$= \frac{£5\,000}{3.478} = £1\,437$$

Average = £1 166

Full net rental value £17 166

say £17 200 pa

Note

1: The yield of 6 per cent is considered to be an appropriate freehold yield for the shop.

The full net rental value should be zoned to obtain a Zone A rent per m^2.

Using a zoning method (as before) with two 5 m zones and a remainder and 'halving back':

$$
\begin{aligned}
\text{Let Zone A per m}^2 &= £X \\
10 \times 5 \times X &= 50X \\
10 \times 5 \times \tfrac{1}{2}X &= 25X \\
10 \times 8 \times \tfrac{1}{4}X &= 20X \\
\hline
&\quad 95X
\end{aligned}
$$

$$X = \frac{£17\,200}{95} = £181 \text{ per m}^2$$

Applying the results to the subject shop:

$$12 \times 5 \times 181 \qquad = £10\,860$$
$$12 \times 5 \times \ 90.5 \qquad = \ £5\,430$$
$$12 \times 10 \times 45.25 \quad = \ £5\,430$$

Full net rental value = £21 720

say   £21 700 pa

The freehold and leasehold interests may now be valued.

Freehold

Shop

Rent received	£18 000 pa	
Less External repairs and insurance	£2 170 pa	
Net income	£15 830 pa	
YP 4 years at 5 per cent [see note 1]	3.546	
		£56 133
Reversion to full net rental value	£21 700 pa	
YP in perpetuity deferred 4 years at		
6 per cent	13.202	£286 483

Offices

Full net rental value [see note 2]	£8 400 pa	
YP in perpetuity at 7 per cent [see note 3]	14.286	£120 002
Capital value		£462 618
say		£462 600

Notes
1: The yield for the 4 year term is 1 per cent less than the reversionary income, as it is considered to be more secure.
2: 240 m^2 net × £35 = £8 400 per annum.
3: 7 per cent is an acceptable freehold initial yield for offices.

Leasehold

Full rental value on internal repairing terms [see note 1]	£23 870 pa
Less Rent paid	£18 000 pa
Profit rent	£5 870 pa
YP 4 years at 7 per cent and 3 per cent	2.135
(tax 40 per cent)	£12 532
Plus 3 × apportioned rateable value [see note 2]	£43 256
	£55 788
Payment to lessee say	£55 800

Notes

1: The full rental value on internal repairing terms is net rental value + external repairs and insurance = £21 700 + £2 170 = £23 870 per annum.

2: The lessee would expect to receive his entitlement to compensation under *Landlord and Tenant Act 1954*. This would be 3 × Apportioned rateable value i.e.

$$\frac{\text{Rental value of occupied portion}}{\text{Total rental value}} = 3 \times \frac{21\,700}{30\,100} \times £20\,000 = £43\,256$$

Note

As in question 3, all parties involved might negotiate for amounts in excess of those calculated. They would hope to receive some proportion of the marriage gain realised when the land is available for the indoor training centre and social club. Because of insufficient information, this is incalculable.

Question 1.5

The first step in answering this question is to analyse transactions B and C to obtain a Zone A rent per m^2 for the shops and an initial freehold yield.

The analysis of transaction B may give a Zone A rent per m^2 after the full net rental value is calculated from adding the 'average' annual equivalent of the premium (as in previous questions) to the rent passing. The weakness of this analysis is the need to assume appropriate yields for freehold and leasehold when working out the annual equivalents.

i.e.

Rent passing	£16 500 pa
Plus Annual equivalent of premium	

$$\text{Freeholder} \quad \frac{£10\,000}{\text{YP 5 years at 6 per cent}} \quad [\text{see note 1}]$$

$$= \frac{£10\,000}{4.212} = £2374$$

$$\text{Lessee} \quad \frac{£10\,000}{\substack{\text{YP 5 years at 7 per cent} \\ \text{and 3 per cent} \\ (\text{tax 40 per cent})}}$$

$$\frac{£10\,000}{2.605} = £3838$$

Average	= £3100 pa
Full net rental value	= £19 600 pa

Note

1: The yield of 6 per cent is considered appropriate for the freehold interest of the shop.

The full net rental value of £19 600 per annum may be zoned to obtain a Zone A rent per m² using two 5 m zones and a remainder and 'halving back' as in previous examples.

$$\text{Let Zone A rent per m}^2 = £X$$

$$
\begin{array}{ll}
9 \times 5 \times X & = 45X \\
9 \times 5 \times \frac{1}{2}X & = 22.5X \\
9 \times 8 \times \frac{1}{4}X & = 18X \\
\text{Offices } 150 \times 0.3X & = 45X \\
\hline
& 130.5X
\end{array}
$$

$$X = \frac{19\,600}{130.5} = £150.19$$

say £150 per m²

$$
\begin{array}{ll}
\text{So that Zone A rents} & = £150 \text{ per m}^2 \\
\text{and office rents} & = £45 \text{ per m}^2
\end{array}
$$

These rents per m² should now be applied to transaction C to give a full net rental value and a freehold initial yield.

Again, using a zoning method as in transaction B:

$$
\begin{array}{lll}
10 \times 5 \times £150 & = & £7\,500 \\
10 \times 5 \times £75 & = & £3\,750 \\
10 \times 5 \times £37.5 & = & £1\,875 \\
\text{Offices } 120 \times £45 & = & £5\,400 \\
\hline
\text{Full net rental value} & = & £18\,525
\end{array}
$$

Initial yield =

$$\frac{\text{Full net rental value}}{\text{Capital value}} = \frac{18\,525}{308\,750}$$

= 6 per cent

This is consistent with the yield assumed in the analysis of transaction B.

Applying the results to transaction A, the net rental value may again be calculated by using a zoning method:

$$
\begin{array}{lll}
8 \times 5 \times £150 & = & £6\,000 \\
8 \times 5 \times £75 & = & £3\,000 \\
8 \times 6 \times £37.5 & = & £1\,800 \\
\text{Offices } 100 \text{ m}^2 \times £45 & = & £4\,500 \\
\hline
\text{Full net rental value} & = & £15\,300 \text{ pa}
\end{array}
$$

Having analysed the transactions of B and C and applied the results to A, it is now necessary to deal with the surrender and renewal of the lease of shop A.

The lessee wishes to surrender a 2 year unexpired term and replace it with a 10 year term with a 5 year review to current market value. To obtain a rent for the first 5 years, a surrender and renewal calculation is carried out, based on the principle of present interest = proposed interest.

Let rent reserved for the 5 year term on full repairing and insuring terms = £X.

Landlord

Present interest

Rent passing	£10 000 pa	
YP 2 years at 5 per cent [see note 1]	1.859	£18 590
Reversion to full net rental value	£15 300 pa	
YP in perpetuity deferred 2 years		
at 6 per cent	14.833	£226 945
	Capital value	£245 535
	say	£245 500

Note

1: It is assumed that the income receivable for the 2 year term is more secure than the reversion so that the initial yield is lowered by 1 per cent.

An alternative approach to the present interest would be to assume that in 2 years' time the freeholder would improve the property at the cost of £20 000 to give a 10 per cent return, i.e. a rental increase of £2 000 per annum, i.e.

Present interest

Term as before		£18 590
Reversion to improved rental value	£17 300 pa	
YP in perpetuity deferred 2 years		
at 6 per cent	14.833	£256 610
		£275 200
Less Cost of improvement	£20 000	
Present value of £1 in 2 years at		
6 per cent	0.89	£17 800
	Capital value	£257 400

It may be advisable to take the higher capital value, i.e. £257 400, to set against the proposed interest.

Proposed interest

Rent passing	£X	
YP 5 years at 5 per cent	4.329	£4.329X
Reversion to improved rental value [see note 1]	£17 300 pa	
YP in perpetuity deferred 5 years at 6 per cent	12.454	£215 454
Capital value		£4.329X + 215 454

$$
\begin{aligned}
\text{Present} &= \text{Proposed} \\
£257\,400 &= £4.329X + 215\,454 \\
X &= £9\,689 \\
\text{say} &= £9\,700 \text{ per annum}
\end{aligned}
$$

Note

1: It is assumed that in 5 years' time the rent would be reviewed to the improved rental value. The '21 year rule' will not apply as the improvement was a condition of the lease.

Lessee

Present interest

Full net rental value	£15 300 pa
Less Rent paid	£10 000 pa
Profit rent	£5 300 pa
YP 2 years at 8 per cent and 3 per cent (tax 40 per cent)	1.11
Capital value	£5 883

Proposed interest

Improved net rental value	£17 300 pa
Less Rent paid	£X
Profit rent	£17 300 − X
YP 5 years at 7 per cent and 3 per cent (tax 40 per cent) [see note 1]	2.605
	£45 066 − 2.605X
Less Cost of improvements	£20 000
Capital value	£25 066 − 2.605X

Note

1: The leasehold yield of 8 per cent has been lowered by 1 per cent to reflect the security of being guaranteed a second 5-year term.

$$\text{Present} = \text{Proposed}$$
$$£5\,883 = £25\,066 - 2.605X$$
$$X = £7\,364$$
$$\text{say} \quad £7\,350 \text{ per annum}$$

Acting between the parties, averaging a rent between £9 700 and £7 350 per annum, recommend a rent of say £8 500 per annum.

2 Investment Appraisal

Investment appraisal is concerned with the evaluation not only of property investments, but also of opportunities in alternative investment media. Often, it involves
a comparison of the two.

Because of this, professional advisers in other disciplines, notably accountants,
have become increasingly involved in investment appraisal and accounting methods
have been absorbed into the evaluation process. This might also be construed as a
reflection on the adequacy of traditional valuation techniques. It is therefore essential that the student of advanced valuation obtains a knowledge of modern appraisal
techniques, and is able to apply them, if the valuer's role in investment appraisal is
not to be further eroded.

The questions in this section begin with general considerations of the various
approaches to investment appraisal, dealing firstly with the simple Payback and
Accounting Rate of Return methods. Discussion then progresses to the more
sophisticated methods, of which discounting theory forms the basis, that is, Net
Present Value, Internal Rate of Return, Incremental Analysis and Yield to Equity.
Later questions demonstrate the application of these to specific circumstances.

An individual or organisation with capital to invest may require advice regarding
the relative merits of available investments. Alternatively, an investor may need to
know whether the existing portfolio is satisfactory, or whether capital could be
more advantageously invested elsewhere. Questions involving both of these situations are presented, demonstrating the principles applied in the comparison of
investments.

Consideration is also given to the effects of taxation, since the payment of tax
must reduce the return from an investment. This again requires individual treatment,
since the implications for each investor will vary according to their tax position.

Investment appraisal is therefore not so much concerned with market valuation,
but rather the evaluation of particular opportunities for specific investors.

INVESTMENT APPRAISAL – QUESTIONS

2.1. Construct examples to illustrate the various techniques that may be employed
in the appraisal and ranking of investments, discussing the problems which may
be encountered in the application of these techniques.

32

2.2. 'As more money is borrowed to finance an investment, the rate of return on the equity element will change, whereas the net present value will remain constant.'

Discuss the validity of this statement, constructing examples to illustrate your answer.

2.3. (a) Appraise the cash flows of the two investments outlined below, by the following methods:

   (i) net present value, assuming a weighted average cost of capital of 15 per cent,

   (ii) internal rate of return,

   (iii) incremental analysis, including the calculation of the incremental internal rate of return.

	Investment A			*Investment B*	
Year	*Outflow*	*Inflow*	*Year*	*Outflow*	*Inflow*
0	£80 000	0	0	£120 000	0
1		£16 000	1		£20 000
2		£28 000	2		£30 000
3		£36 000	3		£42 000
4		£48 000	4		£50 000
			5		£56 000

(b) Discuss the significance of the results of the appraisals carried out in part (a).

2.4. (i) Your client has £75 000 to invest for a period of 4 years. The following 3 investments are available to him:

   (a) the deposit of all, or part, of the money with a bank offering a fixed 15 per cent return per annum,

   (b) the outlay of £52 500 which will produce an income of £15 180, £18 000, £21 000 and £27 000 at the end of the first, second, third and fourth years respectively,

   (c) the outlay of £75 000 which will produce an income of £17 100, £27 000, £33 000 and £34 500 at the end of the first, second, third and fourth years respectively.

Carry out calculations in order to advise your client which investment, or combination of investments, will give the best return on his available capital.

It may be assumed that the internal rate of return of investment (c) is 16 per cent.

(ii) If your client were to deposit his money in a bank, his original capital would be returnable after 4 years and there would, therefore, be no need for the provision for a sinking fund.

Explain, and illustrate, why it would also be unnecessary to provide for a sinking fund to recoup his capital if he were to invest in (b) or (c).

2.5. The sum of £60 000 can be invested in two alternative ways:

(i) As a loan to be repaid over 5 years by annual instalments representing capital, plus interest on the outstanding balance. The annual interest rate is to be fixed at 10 per cent for the first year, 12 per cent for the second and third years and 14 per cent thereafter.

(ii) In a leasehold property with 5 years unexpired, producing net annual income of £16 500 for the first two years and £30 500 thereafter.

Compare the performance of the two investments, by calculating the internal rate of return of each, having first allowed for the payment of income tax at a rate of 40 per cent.

2.6. (i) Discuss the problems which may be encountered when using net present value and internal rate of return to appraise alternative projects and calculate the internal rate of return of each of the following investments:

Project A Cost £200 000; expected income flow £21 400 per annum, fixed in perpetuity, payable annually in advance.

Project B Cost £300 000; expected income flow £20 000 in the first year, rising by 8 per cent per annum, payable annually in arrear, with an assumed capital value of £440 000 at the end of year 5.

Project C Cost £500 000; expected income flow of £40 000 per annum for the first 2 years and £44 000 per annum for the following 3 years, payable annually in advance, with an expected capital value of £640 000 in 5 years' time.

(ii) Assuming your client has £500 000 to invest, advise him which investment, or combination of investments, to choose, giving reasons.

2.7. Seven years ago, a freehold property was purchased with a loan of £80 000. Repayments on the loan were calculated over 15 years, by equal instalments representing both principal and interest, charged at 12 per cent on the outstanding balance.

An offer of £175 000 has just been received for the property. The freeholder therefore has to decide between

(i) accepting the offer and reinvesting his equity elsewhere and

(ii) leaving his equity in the existing investment.

To assist the freeholder in his decision, calculate the yield to redemption from his existing investment, ignoring tax implications.

The property is occupied by several tenants on different leases, producing the following estimated income flow (in current terms) over the next 8 years:

Year	1	2	3	4	5	6	7	8
Income	£12 000	£14 000	£14 000	£18 000	£18 000	£20 000	£20 000	£22 000

INVESTMENT APPRAISAL – SUGGESTED ANSWERS

Question 2.1

The appraisal techniques to consider in answering this question, are as follows:
  (i) Accounting rate of return
 (ii) Payback
(iii) Net present value
 (iv) Internal rate of return
  (v) Incremental analysis
 (vi) Yield to equity.

A brief description of each method should be followed by a simple example and a consideration of the merits, and any disadvantages of the method.

(i) *Accounting rate of return*

This is sometimes called the return on capital employed and is simply the profit produced by an investment, expressed as a percentage of capital outlay.

One method of calculating the accounting rate of return is to express the average annual return from the investment over its life, as a percentage of the average capital value.

Example 1

An investment is purchased for £60 000 and it is estimated that it will be sold 5 years later for £100 000. Income for the first 2 years will be £10 000 per annum and for the following 3 years, £25 000 per annum.

$$\text{average capital value} = \frac{£60\,000 + £100\,000}{2} = £80\,000$$

$$\text{average annual return} = \frac{£10\,000 + £10\,000 + £25\,000 + £25\,000 + £25\,000}{5}$$

$$= £19\,000$$

$$\text{Accounting rate of return} = \frac{£19\,000}{£80\,000} \times 100 = 23.75 \text{ per cent}$$

Accounting rate of return may be used as a guide in the investment decision, comparing the return achieved by the investment with the investor's target return. In the ranking of investments, the preferred investment will be that with the highest accounting rate of return.

The main advantage of the method is its simplicity — it is both simple to understand and to calculate.

The disadvantages to consider are:

(a) There are several ways of calculating the accounting rate of return, therefore the results are open to misinterpretation unless the basis of calculation is clear. For instance, initial capital outlay is sometimes used instead of average capital value. In Example 1, this would result in a higher accounting rate of return:

$$\frac{£19\,000}{£60\,000} \times 100 = 31.67 \text{ per cent}$$

However, this might be considered a more realistic basis, since estimation of future disposal value may be liable to error.

(b) Investments are compared on the basis of their values expressed as percentages, which do not disclose other vital elements of the investment without further investigation. One would, for example, need to consider the initial capital outlay required by alternative investments. An important factor will be the duration of the investment. The investor may only wish to tie up capital for 4 years, but neither the 23.75 per cent nor the 31.67 per cent, calculated above, disclose the fact that these returns are achieved by an investment of 5 years duration.

(c) A very important point is that the timing of income and outlay is ignored. In Example 1, the £25 000 receivable in 5 years' time is considered on exactly the same terms as the £25 000 receivable in 4 years' time and also that in 3 years' time. The present value of these sums would not be the same, but no allowance is made for this fact in the calculation.

(ii) *Payback*

This is another simple method of investment appraisal, which determines the time taken for the income flowing into an investment to recoup the capital outlay.
     Example 1 is again used to illustrate this method. It is always advisable, in answering an examination question, to utilise examples already constructed whereever possible, in order to save time.

Year	Outflow	Inflow	
0	£60 000	£10 000	
1		£10 000	
2		£25 000	
3		£25 000	
4		£25 000	Payback period
5		£25 000)	
		£100 000)	

     By the end of year 4, total income of £70 000 will have been received, therefore the investment would payback the outlay within 4 years.
     Using this method, an investor would decide to accept or reject an investment

depending upon whether or not it will payback within the desired time. Ranking of investments would be on the basis of their comparative lengths of payback, the preferred investment being that with the shortest payback time.

This method is easy to understand and calculation is simple.

The disadvantages to consider are:

(a) Calculation of payback time is not always consistent. Incorrect decisions may result from differing interpretations of the point at which payback begins. If the criterion for accepting an investment is that it should pay back in 5 years, there must be clear understanding of exactly when this 5 year period starts. For instance, does payback commence from the payment of the initial capital outlay or is it when payment of total capital outlay is completed? An alternative interpretation might be that payback commences on receipt of the first income payment.

(b) The choice of payback time is critical. Since the acceptance or rejection of an investment is based upon its ability to meet this criterion, it is obviously of fundamental importance that the length of payback is correctly chosen. An apparently small error of 1 year at this stage, could result in an incorrect investment decision.

(c) The method attaches no importance to later income inflows.

One argument used in support of this method is that by employing the payback period as the decision factor, only the earlier, less risky, income flows are considered. However, to disregard future income flows may ultimately be to disregard the better investment.

Example 2

Two investment opportunities are available, both requiring an initial capital outlay of £100 000.

Income from the investments is as follows:

Investment A   £25 000 per annum for the first 4 years, £30 000 per annum for the following 6 years.

Investment B   £10 000 per annum for the first 4 years, £60 000 per annum for the following 6 years.

On the payback criteria, A would be the preferred investment, payback being achieved in 4 years, compared with the 5 years taken by Investment B. However, future income flows suggest that if the total lives of the investments are considered, B is superior. Over the 10 years, B produces a total income of £400 000, whereas A achieves only £280 000, but the advantages of B do not accrue until after year 4 and would therefore be ignored.

(d) A major disadvantage shared with the accounting rate of return method is that the timing of cash flows is ignored. The flows are considered in terms of their actual amounts, no account being taken of their present values.

(iii) *Net Present Value (NPV)*

Using this method of appraisal, an investment is judged according to whether or not
the income from it balances or outweighs the capital outlay. If it does, the investment is acceptable, if it does not, the investment will be rejected. In this case,
however, allowance is made for the timing of cash flows and each is considered in
terms of its present value, discounted for the necessary period of time at the rate of
return required by the investor. The total of the present values of all outflows is
deducted from the total of the present values of all inflows, the resulting figure
being referred to as net present value (NPV). If the NPV is zero or positive, the
investment is acceptable, since it will achieve or outperform the investor's required
(or target) rate of return. If the NPV is negative, the investment is rejected, since
the target return is not achieved.

When used to rank investments, the one with the highest NPV at the target
return will be the preferred investment.

Example 3

Two investment opportunities are available, with expected income flows as follows:

Year	Investment A	Investment B
0		
1	£7 000	£6 000
2	£8 000	£6 000
3	£9 000	£9 000
4	£10 000	£9 000
5	£11 000	£12 000

Investment A may be purchased for £125 000 and Investment B for £120 000. In
5 years' time, the investments are expected to sell for £175 000 and £180 000
respectively.

The investor requires a 12 per cent return on capital.

Calculation of NPV of Investment A at a discount rate of 12 per cent.

Year	Inflow £	Outflow £	PV of £1 at 12 per cent	PV of inflows £	PV of outflows £
0		125 000	1		125 000[a]
1	7 000		0.893	6 251[b]	
2	8 000		0.797	6 376	
3	9 000		0.712	6 408	
4	10 000		0.636	6 360	
5	11 000)				
	175 000)[c]		0.567	105 462	
		Present value of inflows		130 857	
	less	Present value of outflows		125 000	
		NPV at 12 per cent		+5 857	

Notes

[a] Outlay of £125 000 is immediate, therefore its present value is also £125 000.
[b] £7 000 × present value of £1 in 1 year at 12 per cent = £6 251.
[c] Anticipated sale price at the end of 5 years.

Calculation of NPV of Investment B at a discount rate of 12 per cent.

Year	Inflow £	Outflow £	PV of £1 at 12 per cent	PV of inflows £	PV of outflows £
0		120 000	1		120 000
1	6 000		0.893	5 358	
2	6 000		0.797	4 782	
3	9 000		0.712	6 408	
4	9 000		0.636	5 724	
5	12 000)				
	180 000)		0.567	108 864	
		Present value of inflows		131 136	
	less	Present value of outflows		120 000	
		NPV at 12 per cent		+11 136	

Both investments achieve the required 12 per cent return, but if the investor has to make a choice, Investment B would be preferred, since it has the highest NPV at 12 per cent.

Further consideration is given to NPV in (v) below.

(iv) *Internal rate of return (IRR)*

The internal rate of return is the actual return obtained from an investment; it is the rate of return at which the NPV is zero, the investment making neither a profit nor a loss

In Example 3, at a target rate of 12 per cent, the NPV of Investment A was found to be £5 857 and of Investment B £11 136, which means that the actual rate of return, the IRR, is higher than 12 per cent in both cases.

An investment is acceptable if the IRR is higher than the target rate of return, which applies to both Investment A and Investment B.

Where there are several investment opportunities, those investments with an IRR below the target return will be rejected; of those with an IRR above the target return, the investment with the highest IRR will be preferred.

The IRR may be discovered by linear interpolation between two rates of interest which produce a negative and a positive NPV.

In Example 3, having obtained a positive NPV at a discount rate of 12 per cent, the calculation is repeated at a higher discount rate, in order to achieve a negative NPV. The rate at which the NPV is zero — the IRR — must therefore fall between these two discount rates.

A test rate of 15 per cent will now be applied to Investments A and B.

Calculation of NPV of Investment B at a discount rate of 15 per cent

Year	Inflow £	Outflow £	PV of £1 at 15 per cent	PV of inflows £	PV of outflows £
0		125 000	1		125 000
1	7 000		0.870	6 090	
2	8 000		0.756	6 048	
3	9 000		0.658	5 922	
4	10 000		0.572	5 720	
5	11 000 } 175 000 }		0.497	92 442	
	Present value of inflows			116 222	
less	Present value of outflows			125 000	
	NPV at 15 per cent			− 8 778	

Calculation of NPV of Investment B at a discount rate of 15 per cent.

Year	Inflow £	Outflow £	PV of £1 at 15 per cent	PV of inflows £	PV of outflows £
0		120 000	1		120 000
1	6 000		0.870	5 220	
2	6 000		0.756	4 536	
3	9 000		0.658	5 922	
4	9 000		0.572	5 148	
5	12 000 ⎱ 180 000 ⎰		0.497	95 424	

Present value of inflows	116 250
less Present value of outflows	120 000
NPV at 15 per cent	− 3 750

For both investments the NPV is now available at two discount rates — a positive NPV at 12 per cent and a negative NPV at 15 per cent. The IRR, that is the discount rate at which the NPV is zero, must, in both cases, lie between 12 per cent and 15 per cent. This may be shown diagrammatically.

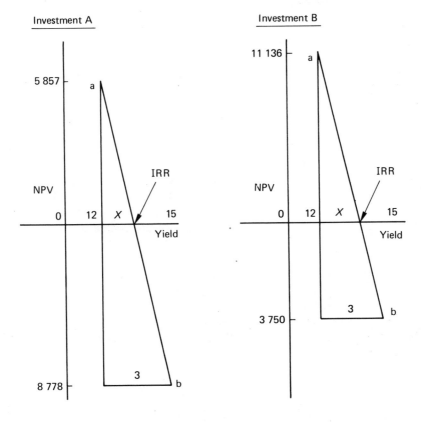

If these graphs were to be drawn accurately, the points at which the NPV is zero could be measured, but this is not necessary, since the principle of similar triangles may be applied to find the length of X in each case.

Investment A

$$\frac{X}{5\,857} = \frac{3}{5\,857 + 8\,778}$$

$$14\,635X = 17\,571$$

$$X = 1.201$$

The IRR of Investment A is thus:

12 per cent + 1.201 per cent = 13.201 per cent

Investment B

$$\frac{X}{11\,136} = \frac{3}{11\,136 + 3\,750}$$

$$14\,886X = 33\,408$$

$$X = 2.244$$

The IRR of Investment B is thus:

12 per cent + 2.244 per cent = 14.244 per cent

These figures are not strictly accurate, since the hypotenuse, ab, is in reality slightly curved, and to obtain a more precise figure, the two test discount rates should be as close together as possible. Ideally, the NPV of Investment A should now be recalculated at 13.201 per cent and of Investment B at 14.244 per cent, but, in an examination, time would preclude this and it would be sufficient merely to note this point regarding accuracy.

The results achieved by IRR analysis confirm those obtained by the NPV method. Calculating the NPV of Investments A and B revealed that both achieved the target return of 12 per cent, with Investment B preferred, since it had the higher NPV. Investment B also has a higher IRR and, using this criteria, would also be preferred.

IRR is given further consideration in (v), below.

(v) *Incremental analysis*

NPV and IRR analysis have so far produced consistency in indicating the preferred investment, but such convenient agreement does not always emerge. Unfortunately, conflict can arise between NPV and IRR analysis and incremental analysis is useful in such a situation.

Example 4

Assume two investment opportunities with the following cash flows:

	Investment A			Investment B	
Year	Outflow	Inflow	Year	Outflow	Inflow
0	£25 000		0	£37 500	
1		£7 500	1		£10 000
2		£10 000	2		£12 500
3		£12 500	3		£12 500
4		£12 500	4		£15 000
			5		£17 500

The NPV of the two investments will first of all be calculated at an assumed 18 per cent target rate of return.

NPV of Investment A at 18 per cent

Year	Inflow £	Outflow £	PV of £1 at 18 per cent	PV of inflows £	PV of outflows £
0		25 000	1		25 000
1	7 500		0.847	6 353	
2	10 000		0.718	7 180	
3	12 500		0.609	7 613	
4	12 500		0.516	6 450	
		Present value of inflows		27 596	
	less	Present value of outflows		25 000	
		NPV at 18 per cent		+ 2 596	

NPV of Investment B at 18 per cent

Year	Inflow £	Outflow £	PV of £1 at 18 per cent	PV of inflows £	PV of outflows £
0		37 500	1		37 500
1	10 000		0.847	8 470	
2	12 500		0.718	8 975	
3	12 500		0.609	7 613	
4	15 000		0.516	7 740	
5	17 500		0.437	7 648	
		Present value of inflows		40 446	
	less	Present value of outflows		37 500	
		NPV at 18 per cent		+ 2 946	

Both investments achieve the required 18 per cent, but, on the NPV criteria, B, with the higher NPV would be the preferred investment.

The IRR of the two investments should now be compared.

In an attempt to obtain a negative NPV, a discount rate of 23 per cent is applied. It is advisable, to save time under examination conditions, to choose a second test rate that is reasonably far removed from the first, so as to increase the chance of achieving a negative NPV, rather than testing, in this case at say 20 per cent, only to find the NPV is still positive and a further calculation is necessary. This comment applies only to examination questions and the remarks in (iv) concerning accuracy of calculations, would apply in any other circumstances.

NPV of Investment A at 23 per cent

Year	Inflow £	Outflow £	PV of £1 at 23 per cent	PV of inflows £	PV of outflows £
0		25 000	1		25 000
1	7 500		0.813	6 098	
2	10 000		0.661	6 610	
3	12 500		0.537	6 713	
4	12 500		0.437	5 463	
		Present value of inflows		24 884	
	less	Present value of outflows		25 000	
		NPV at 23 per cent		− 116	

NPV of Investment B at 23 per cent

Year	Inflow £	Outflow £	PV of £1 at 23 per cent	PV of inflows £	PV of outflows £
0		37 500	1		37 500
1	10 000		0.813	8 130	
2	12 500		0.661	8 263	
3	12 500		0.537	6 713	
4	15 000		0.437	6 555	
5	17 500		0.355	6 213	
		Present value of inflows		35 874	
	less	Present value of outflows		37 500	
		NPV at 23 per cent		− 1 626	

Similar triangles may then be used to discover the IRR of the two investments.

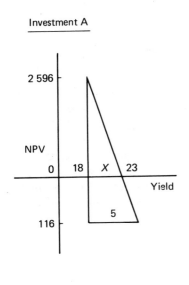

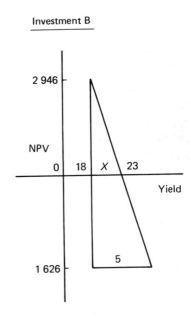

IRR of Investment A

$$\frac{X}{2596} = \frac{5}{2712}$$

$$2712X = 12980$$

$$X = 4.786$$

IRR of Investment A = 18 per cent + 4.786 per cent = 22.786 per cent.

IRR of Investment B

$$\frac{X}{2946} = \frac{5}{4572}$$

$$4572X = 14730$$

$$X = 3.222$$

IRR of Investment B = 18 per cent + 3.222 per cent = 21.222 per cent.

Summary of results

NPV at 18 per cent	Investment A + 2 596	
	Investment B + 2 946	
IRR	Investment A	22.786 per cent
	Investment B	21.222 per cent

There is obviously a conflict between the solutions provided by NPV and IRR.

On the NPV criteria, although both investments achieve the target rate of 18 per cent, Investment B is preferred since at this discount rate it has the higher NPV.

On the IRR criteria, Investment A is preferred, because it has the higher IRR.

If the incremental cash flow between the two investments is evaluated, this will assist in the determination of the preferred investment. It will also disclose the point at which the investor will be indifferent between the two investments.

The incremental IRR may be determined either from the incremental cash flow A–B or B–A. Both are shown here, although in an examination, only one would be necessary.

NPV of incremental cash flow A–B and B–A at 18 per cent

Year	Incremental cash flow A–B[a] £	Incremental cash flow B–A[b] £	PV of £1 at 18 per cent	PV A–B £	PV B–A £
0	+12 500	−12 500	1	+12 500	−12 500
1	− 2 500	+ 2 500	0.847	− 2 118	+ 2 118
2	− 2 500	+ 2 500	0.718	− 1 795	+ 1 795
3	0	0	0.609	0	0
4	− 2 500	+ 2 500	0.516	− 1 290	+ 1 290
5	−17 500	+17 500	0.437	− 7 648	+ 7 648
			NPV	− 351	+ 351

NPV of incremental cash flow A–B at 18 per cent = −351
NPV of incremental cash flow B–A at 18 per cent = +351

Notes
[a] Income or outlay from B deducted from income or outlay from A.
[b] Income or outlay from A deducted from income or outlay from B.

NPV of incremental cash flow A–B and B–A at 23 per cent

Year	Incremental cash flow A–B. £	Incremental cash flow B–A. £	PV of £1 at 23 per cent	PV A–B £	PV B–A £
0	+12 500	−12 500	1	+12 500	−12 500
1	− 2 500	+ 2 500	0.813	− 2 033	+ 2 033
2	− 2 500	+ 2 500	0.661	− 1 653	+ 1 653
3	0	0	0.537	0	0
4	− 2 500	+ 2 500	0.437	− 1 093	+ 1 093
5	−17 500	+17 500	0.355	− 6 213	+ 6 213
			NPV	+ 1 508	− 1 508

NPV of incremental cash flow A–B at 23 per cent = + 1 508

NPV of incremental cash flow B–A at 23 per cent = − 1 508

Calculation of incremental IRR using incremental cash flow A–B

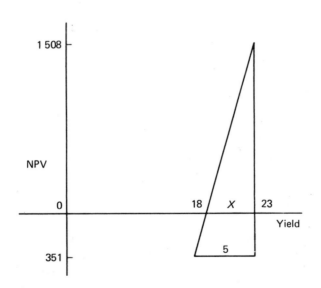

$$\frac{X}{1\,508} = \frac{5}{1\,859}$$

$$1\,859X = 7\,540$$

$$X = 4.056$$

Incremental IRR = 23 per cent − 4.056 per cent = 18.944 per cent.

Advanced Valuation

Calculation of incremental IRR using incremental cashflow B–A

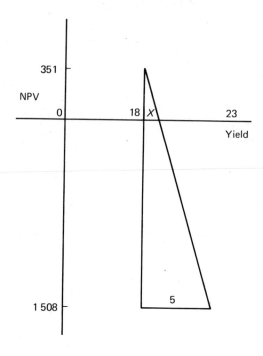

$$\frac{X}{351} = \frac{5}{1\,859}$$

$$1\,859X = 1\,755$$

$$X = 0.944$$

Incremental IRR = 18 per cent + 0.944 per cent = 18.944 per cent.

At the incremental IRR of 18.944 per cent, the NPV of the two investments will be the same. This is often called the 'indifference point', that is, the rate of return at which the investor will be indifferent between the two investments.

A diagram may be used to illustrate the results.

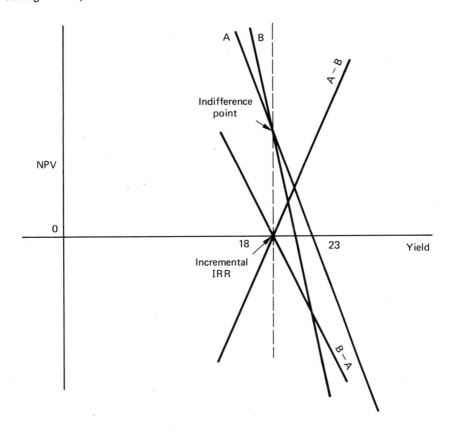

At discount rates below the indifference point (18.944 per cent) Investment B will show a higher NPV than Investment A, whereas at discount rates higher than 18.944 per cent, Investment A will have the higher NPV. Thus, if the NPV criteria is used the decision as to which investment is preferred will depend upon whether the target rate of return is above or below the indifference point.

Alternatively, if the IRR criteria is used, Investment A, with the higher IRR, will always appear to outperform Investment B.

It is often suggested that the investment decision based upon NPV is simple, in that the investment with the highest NPV at a target rate of interest is chosen, whereas the decision is much more complex when based upon IRR, particularly where there is a choice between a number of investments. In such a case, it would be necessary to pair up the various investments and compare incremental cash flows, gradually eliminating investments until the preferred one emerges.

Other problems which may arise in IRR analysis should be noted, including the following:

(a) It is possible for an investment to have more than one IRR. This may occur when the cash flow changes direction, that is a positive cash flow changes to negative, or vice versa. The IRR is likely to change with each change in the direction of the cash flow.

(b) It is also possible for an investment to have an IRR which would render it an acceptable investment, when at all discount rates the NPV is negative.

(c) In some circumstances, it is impossible to calculate the IRR, for example, this may occur when all the money to finance an investment is borrowed.

(vi) *Yield to equity*

This is the yield on the equity portion of capital invested, which varies with the amount of capital that is borrowed.

This would need to be considered in answering the question, but it is not pursued further here, since it is dealt with in more detail in Question **2.2**, to which the reader is referred.

Question 2.2

In the question, the statement regarding return on equity precedes that relating to NPV, but it is suggested that the answer is more easily framed if NPV is dealt with first.

What has to be demonstrated is that the NPV of an investment will remain constant, no matter how much of the original outlay is borrowed.

Example 1

The income from an investment is as follows:

End of year	Income
1	£10 000
2	£15 000
3	£20 000
4	£30 000

The cost of the investment is £50 000 and, initially, it will be assumed that none of this is borrowed.

Assuming a target rate of return of 12 per cent, the NPV of the investment is now calculated.

Year	Inflow £	Outflow £	PV of £1 at 12 per cent	PV of inflows £	PV of outflows £
0		50 000	1		50 000
1	10 000		0.893	8 930	
2	15 000		0.797	11 955	
3	20 000		0.712	14 240	
4	30 000		0.636	19 080	

	Present value of inflows	54 205
less	Present value of outflows	50 000
	NPV at 12 per cent	+ 4 205

The calculation should now be repeated assuming that some proportion of the capital outlay is borrowed. If £10 000 is borrowed, the repayments at the end of each of the 4 years will be

$$\frac{£10\,000}{\text{YP 4 years at 12 per cent}} = \frac{£10\,000}{3.037} = £3\,293$$

Recalculating the NPV will demonstrate that it remains unchanged, even though £10 000 has now been borrowed.

Year	Inflow £	Outflow £	PV of £1 at 12 per cent	PV of inflows £	PV of outflows £
0		40 000	1		40 000
1	10 000	3 293	0.893	8 930	2 940
2	15 000	3 293	0.797	11 955	2 624
3	20 000	3 293	0.712	14 240	2 344
4	30 000	3 293	0.636	19 080	2 094

	Present value of inflows	54 205	50 002
less	Present value of outflows	50 002	
	NPV at 12 per cent	+4203	

There is a rounding error of £2 between the two valuations, but a further calculation will confirm that the results are not accidental.

If £25 000 of the initial outlay is borrowed, repayments at the end of each of the 4 years will be

$$\frac{£25\,000}{\text{YP 4 years at 12 per cent}} = \frac{£25\,000}{3.037} = £8\,232$$

Calculation of the NPV now becomes:

Year	Inflow £	Outflow £	PV of £1 at 12 per cent	PV of inflows £	PV of outflows £
0		25 000	1		25 000
1	10 000	8 232	0.893	8 930	7 351
2	15 000	8 232	0.797	11 955	6 561
3	20 000	8 232	0.712	14 240	5 861
4	30 000	8 232	0.636	19 080	5 235

Present value of inflows	54 205	50 008
less Present value of outflows	50 008	
NPV at 12 per cent	+ 4 197	

Apart from a rounding error, the NPV remains unchanged.

Having considered the NPV of the investment in Example 1, it should now be demonstrated that the yield to equity will change as the amount of money borrowed to finance the investment changes. The equity element is that part of the capital invested that is not borrowed. Yield to equity may be calculated using the following formula:

$$\text{yield to equity} = \frac{\text{IRR} - \left(\begin{array}{c}\text{borrowing} \\ \text{rate}\end{array} \times \begin{array}{c}\text{proportion} \\ \text{borrowed}\end{array}\right)}{\text{proportion of equity}}$$

A glance at this formula reveals that if the borrowing rate is lower than the IRR of the investment, the yield to equity will rise and vice versa. It is also apparent that in order to calculate the yield to equity, it is necessary first of all to determine the IRR.

In Example 1, the NPV of the Investment at 12 per cent was found to be +4 205, which means that the IRR must be more than 12 per cent. A rate of 16 per cent will now be tested, adopting the initial assumption that none of the capital outlay is borrowed.

Year	Inflow £	Outflow £	PV of £1 at 16 per cent	PV of Inflows £	PV of outflows £
0		50 000	1		50 000
1	10 000		0.862	8 620	
2	15 000		0.743	11 145	
3	20 000		0.641	12 820	
4	30 000		0.552	16 560	
		Present value of inflows		49 145	
	less	Present value of outflows		50 000	
		NPV at 16 per cent		− 855	

The IRR of the investment may now be calculated.

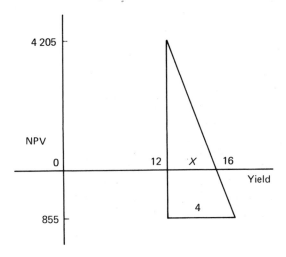

$$\frac{X}{4\,205} = \frac{4}{5\,060}$$

$$5\,060X = 16\,820$$

$$X = 3.324$$

IRR = 12 per cent + 3.324 per cent = 15.324 per cent.

15.324 per cent will also be the yield to equity, since none of the capital outlay is borrowed.

The calculation should now be repeated, assuming that a proportion of the initial outlay is borrowed. If £10 000 is borrowed, repayments at the end of each of the

4 years will be £3 293. Calculation of the NPV at 16 per cent will reveal that the IRR has changed.

Year	Inflow £	Outflow £	PV of £1 at 16 per cent	PV of inflows £	PV of outflows £
0		40 000	1		40 000
1	10 000	3 293	0.862	8 620	2 839
2	15 000	3 293	0.743	11 145	2 447
3	20 000	3 293	0.641	12 820	2 111
4	30 000	3 293	0.552	16 560	1 818

Present value of inflows	49 145	49 215
less Present value of outflows	49 215	
NPV at 16 per cent	− 70	

Calculation of IRR

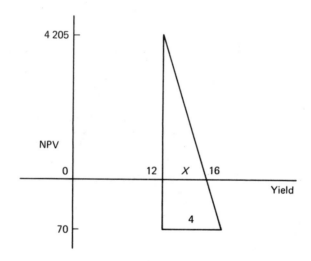

$$\frac{X}{4\,205} = \frac{4}{4\,275}$$

$$4\,275X = 16\,820$$

$$X = 3.935$$

IRR = 12 per cent + 3.935 per cent = 15.935 per cent.

The yield on the equity element may now be calculated.

$$\text{Yield to equity} = \frac{15.935^a - (12^b \times 0.2^c)}{0.8^d}$$

$$= \frac{15.935 - 2.4}{0.8} = 16.919 \text{ per cent}$$

[a] IRR of 15.935 per cent.
[b] Borrowing rate of 12 per cent.
[c] Proportion of capital outlay that is borrowed.
[d] Equity portion of capital outlay.

One further calculation will serve to verify what has been shown so far.

If £25 000 of the outlay is borrowed, repayments will be £8 232 per annum, and the NPV at 16 per cent is as follows:

Year	Inflow £	Outflow £	PV of £1 at 15 per cent	PV of inflows £	PV of outflows £
0		25 000	1		25 000
1	10 000	8 232	0.862	8 620	7 096
2	15 000	8 232	0.743	11 145	6 116
3	20 000	8 232	0.641	12 820	5 277
4	30 000	8 232	0.552	16 560	4 544
	Present value of inflows			49 145	48 033
less	Present value of outflows			48 033	
	NPV at 16 per cent			+ 1 112	

It is obvious that the IRR has increased and must be greater than 16 per cent. A discount rate of 18 per cent is now tested.

Year	Inflow £	Outflow £	PV of £1 at 18 per cent	PV of inflows £	PV of outflows £
0		25 000	1		25 000
1	10 000	8 232	0.847	8 470	6 973
2	15 000	8 232	0.718	10 770	5 911
3	20 000	8 232	0.609	12 180	5 013
4	30 000	8 232	0.516	15 480	4 248
	Present value of inflows			46 900	47 145
less	Present value of outflows			47 145	
	NPV at 18 per cent			− 245	

The IRR must therefore lie betweeen 16 per cent and 18 per cent.

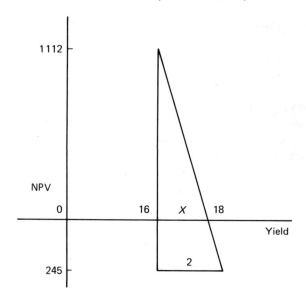

$$\frac{X}{1\,112} = \frac{2}{1\,357}$$

$$1\,357X = 2\,224$$

$$X = 1.639$$

IRR = 16 per cent + 1.639 per cent = 17.639 per cent.

$$\text{Yield to equity} = \frac{17.639 - (12 \times 0.5)}{0.5}$$

$$= 23.278 \text{ per cent}$$

Assuming a borrowing rate of 12 per cent, the results obtained may be summarised.

Proportion borrowed	NPV	IRR	Yield to equity
0	4 205	15.324 per cent	15.324 per cent
20 per cent	4 205	15.935 per cent	16.919 per cent
50 per cent	4 205	17.639 per cent	23.278 per cent

This demonstrates clearly that both the IRR and the yield to equity vary with the amount of money borrowed to finance an investment. Both are seen to increase as the amount borrowed increases. The NPV, on the other hand, is unchanged.

Thus, the statement made in the question is shown to be valid, provided the rate of interest at which capital is borrowed remains constant. The results tend to indicate that NPV is a more reliable criterion than IRR when utilised in the investment decision (see also answer to Question **2.1**). They may also have implications for an investor when deciding what proportion of capital outlay for an investment should be financed by borrowing, since both the IRR and the yield on the equity element increase as borrowing increases.

This seems to indicate that, in order to achieve the optimum return, the investor would be advised to borrow as much of the initial capital outlay as possible. However, this will depend upon the rate of interest at which capital can be borrowed and, although time may prevent this in an examination situation, the effect on the IRR and the yield to equity of a change in the borrowing rate is illustrated below. In fact, if the borrowing rate increases, the IRR and the yield to equity will fall, and vice versa.

When £25 000 was borrowed at an interest rate of 12 per cent, the IRR of the investment in Example 1 was found to be 17.639 per cent, giving a yield to equity of 23.278 per cent.

It will now be assumed that the £25 000 is borrowed at a rate of 18 per cent. Repayments at the end of the 4 years will be

$$\frac{£25\,000}{\text{YP 4 years at 18 per cent}} = \frac{£25\,000}{2.690} = £9\,294$$

In the expectation of finding a lower IRR, a test discount rate of 15 per cent will be applied.

Year	Inflow £	Outflow £	PV of £1 at 15 per cent	PV of inflows £	PV of outflows £
0		25 000	1		25 000
1	10 000	9 294	0.870	8 700	8 086
2	15 000	9 294	0.756	11 340	7 026
3	20 000	9 294	0.658	13 160	6 115
4	30 000	9 294	0.572	17 160	5 316
		Present value of inflows		50 360	51 543
	less	Present value of outflows		51 543	
		NPV at 15 per cent		− 1 183	

The negative NPV indicates that the IRR must be less than 15 per cent and a discount rate of 13 per cent will be tested.

Year	Inflow £	Outflow £	PV of £1 at 13 per cent	PV of inflows £	PV of outflows £
0		25 000	1		25 000
1	10 000	9 294	0.885	8 850	8 225
2	15 000	9 294	0.783	11 745	7 277
3	20 000	9 294	0.693	13 860	6 441
4	30 000	9 294	0.613	18 390	5 697

Present value of inflows	52 845	52 640
less Present value of outflows	52 640	
NPV at 13 per cent	+ 205	

The IRR must therefore lie between 13 per cent and 15 per cent.

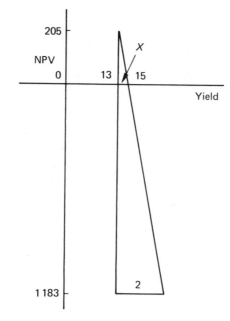

$$\frac{X}{205} = \frac{2}{1\,388}$$

$$1\,388X = 410$$

$$X = 0.295$$

IRR = 13 per cent + 0.295 per cent = 13.295 per cent.

$$\text{Yield to equity} = \frac{13.295 - (18 \times 0.5)}{0.5} = 8.59 \text{ per cent}$$

Thus, because the borrowing rate has risen to 18 per cent, the IRR has fallen by 4.344 per cent and the yield to equity by 14.688 per cent.

The IRR and the yield to equity have therefore been shown to vary not only with the amount of capital that is borrowed to finance an investment, but also with the rate of interest at which that capital is borrowed.

Question 2.3

(a) (i) *Calculation of NPV*

The weighted average cost of capital is given as 15 per cent and the NPV of the two investments should therefore be calculated at a discount rate of 15 per cent.

Investment A

Year	Inflow £	Outflow £	PV of £1 at 15 per cent	PV of inflows £	PV of outflows £
0		80 000	1		80 000
1	16 000		0.870	13 920	
2	28 000		0.756	21 168	
3	36 000		0.658	23 688	
4	48 000		0.572	27 456	
	Present value of inflows			86 232	
	less Present value of outflows			80 000	
	NPV of Investment A at 15 per cent			+ 6 232	

Investment B

Year	Inflow £	Outflow £	PV of £1 at 15 per cent	PV of inflows £	PV of outflows £
0		120 000	1		120 000
1	20 000		0.870	17 400	
2	30 000		0.756	22 680	
3	42 000		0.658	27 636	
4	50 000		0.572	28 600	
5	56 000		0.497	27 832	
	Present value of inflows			124 148	
	less Present value of outflows			120 000	
	NPV of Investment B at 15 per cent			+ 4 148	

(ii) *Calculation of IRR*

From the results in (a) (i), it is clear that both investments achieve a return of 15 per cent, Investment A being preferred, since it produces the highest NPV.

The IRR is the rate of return at which the NPV is zero, therefore it is necessary to calculate the NPV at a higher rate of interest, so as to obtain a negative NPV and then discover the IRR by interpolation.

Extrapolation may be used if the NPV at both discount rates is positive, but interpolation is simpler.

Both investments will be tested at a discount rate of 20 per cent.

Investment A

Year	Inflow £	Outflow £	PV of £1 at 20 per cent	PV of inflows £	PV of outflows £
0		80 000	1		80 000
1	16 000		0.833	13 328	
2	28 000		0.694	19 432	
3	36 000		0.579	20 844	
4	48 000		0.482	23 136	

	Present value of inflows	76 740
less	Present value of outflows	80 000
	NPV of Investment A at 20 per cent	− 3 260

Investment B

Year	Inflow £	Outflow £	PV of £1 at 20 per cent	PV of inflows £	PV of outflows £
0		120 000	1		120 000
1	20 000		0.833	16 660	
2	30 000		0.694	20 820	
3	42 000		0.579	24 318	
4	50 000		0.482	24 100	
5	56 000		0.402	22 512	

	Present value of inflows	108 410
less	Present value of outflows	120 000
	NPV of Investment B at 20 per cent	− 11 590

Having obtained a negative and positive NPV at two discount rates for both Investment A and Investment B, it is now possible to calculate the IRR of the two investments.

Investment A

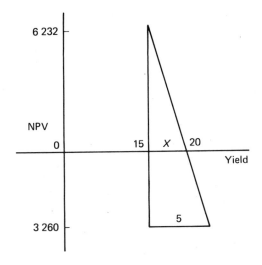

$$\frac{X}{6\,232} = \frac{5}{9\,492}$$

$$9\,492X = 31\,160$$

$$X = 3.283$$

IRR of Investment A = 15 per cent + 3.283 per cent

= 18.283 per cent.

Investment B

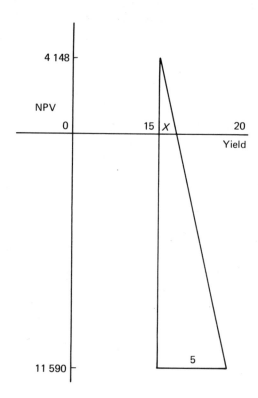

$$\frac{X}{4\,148} = \frac{5}{15\,738}$$

$$15\,738X = 20\,740$$

$$X = 1.318$$

IRR of Investment B = 15 per cent + 1.318 per cent

= 16.318 per cent.

To obtain a more precise IRR, test rates much closer together would be adopted, but under examination conditions, such precision is not possible and is therefore not adopted here.

Using IRR as the appraisal technique, the results obtained by the NPV are confirmed. Once again, A, with the higher IRR, is the preferred investment.

(iii) *Incremental analysis*

This entails consideration of the incremental cash flow between Investment A and Investment B. The calculations may be carried out using either the incremental cash flow A–B or B–A. Only one would be necessary in answering an examination question, but both are shown here to demonstrate the method.

NPV of Incremental cash flow A–B and B–A at a discount rate of 15 per cent

Year	Incremental cash flow A–B[a] £	Incremental cash flow B–A[b] £	PV of £1 at 15 per cent	PV A–B £	PV B–A £
0	+40 000	−40 000	1	+40 000	−40 000
1	−40 00	+4 000	0.870	−3 480	+3 480
2	−2 000	+2 000	0.756	−1 512	+1 512
3	−6 000	+6 000	0.658	−3 948	+3 948
4	−2 000	+2 000	0.572	−1 144	+1 144
5	−56 000	+56 000	0.497	−27 832	+27 832
		NPV		+2 084	−2 084

NPV of incremental cash flow A -B at 15 per cent = +2 084.

NPV of incremental cash flow B–A at 15 per cent = −2 084.

Notes
[a] Income or outlay from B deducted from income or outlay from A.
[b] Income or outlay from A deducted from income or outlay from B.

NPV of incremental cash flow A–B and B–A at a discount rate of 10 per cent

Year	Incremental cash flow A–B £	Incremental cash flow B–A £	PV of £1 at 10 per cent	PV A–B £	PV B–A £
0	+40 000	−40 000	1	+40 000	−40 000
1	−4 000	+4 000	0.909	−3 636	+3 636
2	−2 000	+2 000	0.826	−1 652	+1 652
3	−6 000	+6 000	0.751	−4 506	+4 506
4	−2 000	+2 000	0.683	−1 366	+1 366
5	−56 000	+56 000	0.621	−34 776	+34 776
		NPV		−5 936	+5 936

NPV of incremental cash flow A–B at 10 per cent = −5 936

NPV of incremental cash flow B–A at 10 per cent = +5 936

The IRR of the differential cash flow between the two investments may now be calculated — this is the incremental IRR.

Calculation of incremental IRR using incremental cash flow A–B

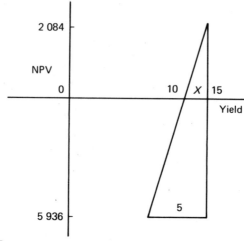

$$\frac{X}{2\,084} = \frac{5}{8\,020}$$

$$8\,020X = 10\,420$$

$$X = 1.299$$

Incremental IRR = 15 per cent − 1.299 per cent

$$= 13.701 \text{ per cent.}$$

Calculation of incremental IRR using incremental cash flow B–A

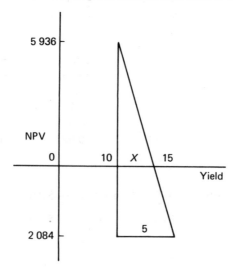

$$\frac{X}{5\,936} = \frac{5}{8\,020}$$

$$8\,020X = 29\,680$$

$$X = 3.701$$

Incremental IRR = 10 per cent + 3.701 per cent

$$= 13.701 \text{ per cent.}$$

(b) The results from part (a) should first of all be summarised.

NPV at 15 per cent: Investment A + £6 232

Investment B + £4 148

A, with the higher NPV, is the preferred investment.

IRR: Investment A   18.283 per cent

Investment B   16.318 per cent

A, with the higher IRR is the preferred investment.

Incremental IRR: 13.701 per cent

Conflict appears to arise between the NPV and IRR criteria, since at discount rates below the incremental IRR, Investment B will have the higher IRR. For example, at a discount rate of 10 per cent, the difference between the NPV of Investment A and that of Investment B is £5 936 in favour of Investment B.

When ranking investments using the NPV criteria, the investment with the highest NPV is preferred. Investment B would therefore be chosen if the target rate of return was below the incremental IRR. However, comparing the investments in the question, A must remain the preferred investment, since the cost of capital is 15 per cent, which is above the incremental IRR.

Incremental analysis is a useful tool in the process of decision between investments, particularly where there is conflict between the results produced by NPV and IRR. More detailed consideration is given to this matter in Question **2.1**.

Question 2.4

(i) First of all it is necessary to determine the IRR of each of the 3 investments.

Investment (a)
No calculation is needed in this case, since the return is fixed at 15 per cent per annum, which is therefore the IRR.

Investment (b)
Calculation of NPV at test rate of 18 per cent

Year	Inflow £	Outflow £	PV of £1 at 18 per cent	PV of inflows £	PV of outflows £
0		52 500	1		52 500
1	15 180	.	0.847	12 857	
2	18 000		0.718	12 924	
3	21 000		0.609	12 789	
4	27 000		0.516	13 932	

Present value of inflows	52 502	
less Present value of outflows	52 500	
NPV at 18 per cent	+ 2	

The IRR must therefore be very close to 18 per cent and, for the purpose of this example, will be assumed to be 18 per cent.

Investment (c)
No calculation is necessary, since the IRR is given as 16 per cent.

It appears that Investment (b) is the better investment. It has an IRR of 18 per cent, compared with the 16 per cent IRR of Investment (c) and the bank deposit (a) with a return of 15 per cent.

Investment (c) also requires the investor to make an additional outlay of £22 500 above Investment (b). If (b) were chosen, what would the investor do with the remaining £22 500? It is a reasonable assumption that £52 500 would be invested in (b) at a return of 18 per cent and £22 500 would be deposited in the bank to obtain a return of 15 per cent. Alternatively, if the investor chooses (c), what rate of return would the £22 500 earn? The IRR of (c) is 16 per cent, but this is the return obtained on the total £75 000. Incremental analysis may be used to solve this dilemma. The incremental IRR will show the yield obtained on the additional outlay of £22 500 required to invest in (c). The NPV of the incremental cash flow will be calculated at 15 per cent, to determine whether the extra outlay will achieve the same return as £22 500 deposited in the bank.

Year	Incremental cash flow c–b £	PV of £1 at 15 per cent	PV c–b £
0	−22 500	1	−22 500
1	+ 1 920	0.870	+ 1 670
2	+ 9 000	0.756	+ 6 804
3	+12 000	0.658	+ 7 896
4	+ 7 500	0.572	+ 4 290
	NPV of incremental cash flow at 15 per cent		− 1 840

The incremental outlay of £22 500 invested in (c) would not earn a 15 per cent return, which confirms the earlier conclusion reached from the IRR calculations, that Investment (b) is preferable.

In fact, the incremental yield is only 11.459 per cent, as illustrated below:

Year	Incremental cash flow c–b £	PV of £1 at 11 per cent	PV c–b £	PV of £1 at 12 per cent	PV c–b £
0	−22 500	1	−22 500	1	−22 500
1	+ 1 920	0.901	+ 1 730	0.893	+ 1 715
2	+ 9 000	0.812	+ 7 308	0.797	+ 7 173
3	+12 000	0.731	+ 8 772	0.712	+ 8 544
4	+ 7 500	0.659	+ 4 943	0.636	+ 4 770
	NPV of incremental cash flow		+ 253		− 298

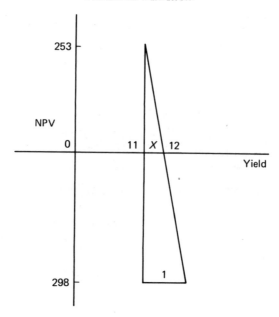

$$\frac{X}{253} = \frac{1}{551}$$

$$551X = 253$$

$$X = 0.459$$

Incremental IRR = 11 per cent + 0.459 per cent

= 11.459 per cent.

Thus, to obtain the best return on his capital, it would be more advantageous for the client to invest £52 500 in Investment (b) and to deposit the remaining £22 500 in the bank.

(ii) The answer should explain that it is not necessary to provide for an annual sinking fund if the client invests in (b) or (c), because the annual income generated by the investments provides not only for interest on capital, but also for recoupment of the capital by the end of 4 years.

This may be illustrated very simply, by showing the addition of interest on the outstanding balance of outlay each year, then deducting the annual income.

At the end of 4 years, it will be seen that the initial outlay has been repaid.

Investment (b)		*Investment (c)*	
Outlay	£52 500	Outlay	£75 000
add Interest at 18 per cent	£ 9 450	*add* Interest at 16 per cent	£12 000
	£61 950		£87 000
less Income	£15 180	*less* Income	£17 100
Outstanding balance		Outstanding balance	
at end of year 1	£46 770	at end of year 1	£69 900
add Interest at 18 per cent	£ 8 419	*add* Interest at 16 per cent	£11 184
	£55 189		£81 084
less Income	£18 000	*less* Income	£27 000
Outstanding balance		Outstanding balance	
at end of year 2	£37 189	at end of year 2	£54 084
add Interest at 18 per cent	£ 6 694	*add* Interest at 16 per cent	£ 8 653
	£43 883		£62 737
less Income	£21 000	*less* Income	£33 000
Outstanding balance		Outstanding balance	
at end of year 3	£22 883	at end of year 3	£29 737
add Interest at 18 per cent	£ 4 119	*add* Interest at 16 per cent	£ 4 758
	£27 002		£34 495
less Income	£27 000	*less* Income	£34 500
	£ 2		£ −5

The initial capital outlay has been repaid — the errors of £2 and £5 being due to rounding.

Question 2.5

(i) In order to find the IRR of this investment, the repayments on the loan must be determined. Since income tax is payable at 40 per cent, it is first of all necessary to calculate the net of tax interest rate payable on the loan.

First year

Interest rate 10 per cent − 40 per cent = 6 per cent net.

Second and third years

Interest rate 12 per cent − 40 per cent = 7.2 per cent net.

Fourth and fifth years

Interest rate 14 per cent − 40 per cent = 8.4 per cent net.

The repayments on the loan of £60 000 may now be calculated.

$$\text{Instalment for first year} = \frac{£60\,000}{\text{YP 5 years at 6 per cent}} = \frac{£60\,000}{4.212}$$

$$= £14\,245$$

The instalment gross of tax will be

$$£14\,245 \times \frac{100}{60} = £23\,742$$

After payment of the first instalment, the amount outstanding on the loan will be

$$£14\,245 \times \text{YP 4 years at 6 per cent}$$

$$= £14\,245 \times 3.465 = £49\,359$$

Instalments in the second and third years

$$= \frac{£49\,359}{\text{YP 4 years at 7.2 per cent}} = \frac{£49\,359}{3.372} = £14\,638$$

The instalments gross of tax will be

$$£14\,638 \times \frac{100}{60} = £24\,397$$

After payment of the second and third instalments, the amount outstanding on the loan will be

$$£14\,638 \times \text{YP 2 years at 7.2 per cent}$$

$$= £14\,638 \times 1.803 = £26\,392$$

Instalments in the fourth and fifth years

$$= \frac{£26\,392}{\text{YP 2 years at 8.4 per cent}} = \frac{£26\,392}{1.774} = £14\,877$$

The instalments gross of tax will be

$$£14\,877 \times \frac{100}{60} = £24\,795$$

Instalments to repay the loan will be flows into the investment. Having calculated these, the IRR of the investment may now be determined. From the net of tax interest rates, it appears that the return may be in the region of 7 per cent and the NPV will first of all be tested at this discount rate.

Year	Inflow £	Tax at 40 per cent £	Net of tax inflow	Outflow £	PV of £1 at 7 per cent	PV of inflows £	PV of outflows £
0				60 000	1		60 000
1	23 742	9 497	14 245		0.935	13 319	
2	24 397	9 759	14 638		0.873	12 779	
3	24 397	9 759	14 638		0.816	11 945	
4	24 795	9 918	14 877		0.763	11 351	
5	24 795	9 918	14 877		0.713	10 607	
		Present value of inflows				60 001	
less		Present value of outflows				60 000	
		NPV at 7 per cent				+ 1	

An NPV of 1 is sufficiently close to nil, so that it may be assumed that the IRR of this investment is 7 per cent

(ii) Since the IRR of the investment in the loan has been established at 7 per cent, the NPV of the leasehold property investment will be tested at a discount rate of 7 per cent.

Year	Inflow £	Tax at 40 per cent £	Net of tax inflow £	Outflow £	PV of £1 at 7 per cent £	PV of inflows £	PV of outflows £
0				60 000	1		60 000
1	16 500	6 600	9 900		0.935	9 257	
2	16 500	6 600	9 900		0.873	8 643	
3	30 500	12 200	18 300		0.816	14 933	
4	30 500	12 200	18 300		0.763	13 963	
5	30 500	12 200	18 300		0.713	13 048	
		Present value of inflows				59 844	
less		Present value of outflows				60 000	
		NPV at 7 per cent				− 156	

Investment in the leasehold property will not quite achieve a 7 per cent return and a discount rate of 6 per cent will be tested.

Net of tax income		£ 9 900 pa	
YP 2 years at 6 per cent [see note 1]		1.833	£18 147
Net of tax income		£18 300 pa	
YP 3 years at 6 per cent [see note 1]	2.673		
x PV of £1 in 2 years at 6 per cent	0.89	2.379	£43 536
Present value of inflows			£61 683
less Present value of outflows			£60 000
NPV at 6 per cent			+ 1 683

Note

1: Where the income is constant over several years, the Years' Purchase may be applied to the income for the appropriate number of years, rather than discounting each year's income separately.

The IRR of the leasehold property must therefore lie between 6 per cent and 7 per cent.

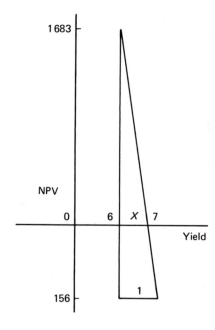

$$\frac{X}{1\,683} = \frac{1}{1\,839}$$

$$1\,839X = 1\,683$$

$$X = 0.915$$

IRR of the leasehold investment = 6 per cent + 0.915 per cent

= 6.915 per cent.

The net of tax IRRs of the two investments are very similar, the loan performing marginally better than the leasehold property. The implication of the question is that IRR will be the criterion upon which the two investments will be compared, but there are other considerations that the investor might be wise to take into account. One of these is security of the income. Is the investor equally sure of receiving the loan repayments as he would be of the rent from the leasehold property? The question does not disclose the landlord's obligations in respect of the leasehold property and the income is quoted as net, but is there likely to be any future additional expenditure?

Although it would appear that the loan is the preferred investment, the investor's target return is not known. If it is say 6.75 per cent, then both investments are acceptable, the loan preferred. But what if the target net of tax return is 5 per cent? In that case, incremental analysis would be advisable, since, at a lower target rate, the leasehold property might produce the higher NPV.

The question does not require this, but the calculation will now be carried out to illustrate the point.

Loan = (i)
Leasehold property = (ii)

Year	Incremental cash flow (i)–(ii) £	Incremental cash flow (ii)–(i) £	PV of £1 at 7 per cent	PV (i)–(ii) £	PV (ii)–(i) £
0	0	0	1	0	0
1	+4 345	−4 345	0.935	+4 063	−4 063
2	+4 738	−4 738	0.873	+4 136	−4 136
3	−3 662	+3 662	0.816	−2 988	+2 988
4	−3 423	+3 423	0.763	−2 612	+2 612
5	−3 423	+3 423	0.713	−2 441	+2 441
		NPV		+ 158	− 158

NPV of incremental cash flow (i)–(ii) at 7 per cent = +158

NPV of incremental cash flow (ii)–(i) at 7 per cent = −158

Year	Incremental cash flow (i)–(ii) £	Incremental cash flow (ii)–(i) £	PV of £1 at 6 per cent	PV (i)–(ii) £	PV (ii)–(i) £
0	0	0	1	0	0
1	+4 345	−4 345	0.943	+4 097	−4 097
2	+4 738	−4 738	0.890	+4 217	−4 217
3	−3 662	+3 662	0.840	−3 076	+3 076
4	−3 423	+3 423	0.792	−2 711	+2 711
5	−3 423	+3 423	0.747	−2 557	+2 557
		NPV		− 30	+ 30

NPV of incremental cash flow (i)–(ii) at 6 per cent = −30

NPV of incremental cash flow (ii)–(i) at 6 per cent = +30

Calculation of incremental IRR using incremental cash flow (i)–(ii)

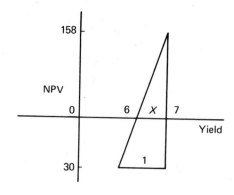

$$\frac{X}{158} = \frac{1}{188}$$

$$188X = 158$$

$$X = 0.840$$

Incremental IRR = 7 per cent − 0.84 per cent

= 6.16 per cent.

Alternatively, the incremental IRR may be calculated using the incremental cash flow (ii)–(i):

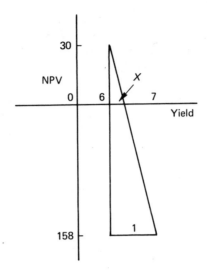

$$\frac{X}{30} = \frac{1}{188}$$

$$188X = 30$$

$$X = 0.16$$

Incremental IRR = 6 per cent + 0.16 per cent

= 6.16 per cent.

At a return of 6.16 per cent, the investor will be indifferent between the two investments, since they will both have the same NPV.

At interest rates below the indifference point, the leasehold property will show the higher NPV, but at interest rates above the indifference point, the loan will show the higher NPV.

Thus, acceptance of the initial results, produced by calculation of the IRR of the two investments, will result in the loan as the preferred investment. However, the IRR decision is very marginal in this case and, if the net of tax target rate of return is 6.16 per cent or below, the investor should be advised seriously to consider preference for the leasehold property investment.

Question 2.6

(i) The discussion required in the first part of this section of the question is not provided here, since the subject is considered in the answers to Questions 2.1 and 2.2, to which the reader is referred.

Calculation of internal rates of return

Project A

A discounted cash flow will be used and it will be assumed, for comparison with Projects B and C, that the investment will be sold in 5 years' time. Since the income is fixed in perpetuity, it may also be assumed that the sale price will be £200 000.

A discount rate of 12 per cent will first of all be tested.

Year	Inflow £	Outflow £	PV of £1 at 12 per cent	PV of inflows £	PV of outflows £
0	21 400	200 000	1	21 400	200 000
1	21 400		0.893	19 110	
2	21 400		0.797	17 056	
3	21 400		0.712	15 237	
4	21 400		0.636	13 610	
5	200 000		0.567	113 400	

	Present value of inflows	199 813
less	Present value of outflows	200 000
	NPV at 12 per cent	−187

The investment does not achieve a return of 12 per cent and a discount rate of 11 per cent will be tested.

Year	Inflow £	Outflow £	PV of £1 at 11 per cent	PV of inflows £	PV of outflows £
0	21 400	200 000	1	21 400	200 000
1	21 400		0.901	19 281	
2	21 400		0.812	17 377	
3	21 400		0.731	15 643	
4	21 400		0.659	14 103	
5	200 000		0.593	118 600	

	Present value of inflows	206 404
less	Present value of outflows	200 000
	NPV at 11 per cent	+ 6 404

The IRR of Project A must therefore lie between 11 per cent and 12 per cent.

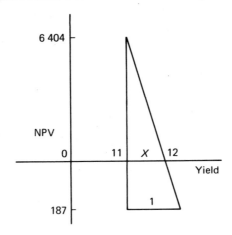

$$\frac{X}{6\,404} = \frac{1}{6\,591}$$

$$6\,591X = 6\,404$$

$$X = 0.972$$

IRR of Project A = 11 per cent + 0.972 per cent

$$= 11.972 \text{ per cent.}$$

The IRR of Project A can, however, be calculated in a much simpler way.

The net outflow is initially £200 000 − £21 400 = £178 600. £21 400 will then be received annually in perpetuity. £178 600 has therefore been paid for a perpetual income of £21 400 per annum, producing a yield of:

$$\frac{£21\,400}{£178\,600} \times 100 = 11.98 \text{ per cent}$$

11.98 per cent will be adopted as the IRR for Project A in the remainder of the answer.

Project B
The income at the end of the first year is £20 000, but this will rise by 8 per cent per annum. Before the IRR can be calculated, the income receivable in the following 4 years must be determined.

Income at end of year 1 = £20 000
Income at end of year 2 = £20 000 + £1 600 = £21 600
Income at end of year 3 = £21 600 + £1 728 = £23 328
Income at end of year 4 = £23 328 + £1 866 = £25 194
Income at end of year 5 = £25 194 + £2 016 = £27 210

Advanced Valuation

Unlike Project A, where the income is fixed, Project B is showing growth. Intuition indicates a higher IRR for Project B and a discount rate of 15 per cent will be tested.

Year	Inflow £	Outflow £	PV of £1 at 15 per cent	PV of inflows £	PV of outflows £
0		300 000	1		300 000
1	20 000		0.870	17 400	
2	21 600		0.756	16 330	
3	23 328		0.658	15 350	
4	25 194		0.572	14 411	
5	27 210 ⎱ 440 000 ⎰		0.497	232 203	
	Present value of inflows			295 694	
	less Present value of outflows			300 000	
	NPV at 15 per cent			−4 306	

The investment does not achieve a return of 15 per cent and a discount rate of 14 per cent will be tested.

Year	Inflow £	Outflow £	PV of £1 at 14 per cent	PV of inflows £	PV of outflows £
0		300 000	1		300 000
1	20 000		0.877	17 540	
2	21 600		0.769	16 610	
3	23 328		0.675	15 746	
4	25 194		0.592	14 915	
5	27 210 ⎱ 440 000 ⎰		0.519	242 482	
	Present value of inflows			307 293	
	less Present value of outflows			300 000	
	NPV at 14 per cent			+ 7 293	

IRR of Project B must therefore lie between 14 per cent and 15 per cent.

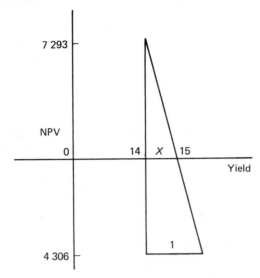

$$\frac{X}{7\,293} = \frac{1}{11\,599}$$

$$11\,599X = 7\,293$$

$$X = 0.629$$

IRR of Project B = 14 per cent + 0.629 per cent

= 14.629 per cent.

Project C
A discount rate of 14 per cent will initially be tested.

Year	Inflow £	Outflow £	PV of £1 at 14 per cent	PV of inflows £	PV of outflows £
0	40 000	500 000	1	40 000	500 000
1	40 000		0.877	35 080	
2	44 000		0.769	33 836	
3	44 000		0.675	29 700	
4	44 000		0.592	26 048	
5	640 000		0.519	332 160	
		Present value of inflows		496 824	
	less	Present value of outflows		500 000	
		NPV at 14 per cent		−3 176	

Advanced Valuation

The investment does not achieve a return of 14 per cent and a discount rate of 13 per cent will be tested.

Year	Inflow £	Outflow £	PV of £1 at 13 per cent	PV of inflows £	PV of outflows £
0	40 000	500 000	1	40 000	500 000
1	40 000		0.885	35 400	
2	44 000		0.783	34 452	
3	44 000		0.693	30 492	
4	44 000		0.613	26 972	
5	640 000		0.543	347 520	

Present value of inflows	514 836
less Present value of outflows	500 000
NPV at 13 per cent	+14 836

The IRR of Project C must therefore lie between 13 per cent and 14 per cent.

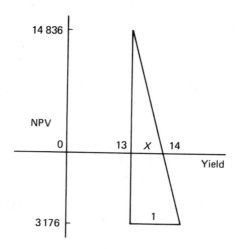

$$\frac{X}{14\,836} = \frac{1}{18\,012}$$

$$18\,012X = 14\,836$$

$$X = 0.824$$

IRR of Project C = 13 per cent + 0.824 per cent

= 13.824 per cent.

An alternative method of calculating the IRR is as follows:

The net outflow is initially £500 000 — £40 000 = £460 000. Future inflows may now be considered as receivable annually in arrears.

Calculation of NPV at 14 per cent

Income		£40 000	
YP 1 year at 14 per cent		0.877	£35 080
Income		£44 000	
YP 3 years at 14 per cent	2.322		
x PV of £1 in 1 year at 14 per cent	0.877	2.036	£89 584
Capital value in 5 years' time		£640 000	
x PV of £1 in 5 years at 14 per cent		0.519	£332 160
Present value of inflows			£456 824
less Present value of outflows			£460 000
NPV at 14 per cent			−3 176

Calculation of NPV at 13 per cent

Income		£40 000	
YP 1 year at 13 per cent		0.885	£35 400
Income		£44 000	
YP 3 years at 13 per cent	2.361		
x PV of £1 in 1 year at 13 per cent	0.885	2.089	£91 916
Capital value in 5 years' time		£640 000	
x PV of £1 in 5 years at 13 per cent		0.543	£347 520
Present value of inflows			£474 836
less Present value of outflows			£460 000
NPV at 13 per cent			+ 14 836

IRR of Project B, as previous calculation, 13.824 per cent.

Summary of results

Project A	*Project B*	*Project C*
(cost £200 000)	(cost £300 000)	(cost £500 000)
IRR 11.98 per cent	IRR 14.629 per cent	IRR 13.824 per cent

(ii) Assuming that the client wishes to invest the entire £500 000, it appears that the alternatives are:

(a) to invest in both A and B

(b) to invest only in C,

since the outlay will be the same in both cases.

It is necessary to show which of these two alternatives will be most favourable to the client, and incremental analysis will assist in the decision.

There are various ways of approaching this and, although only one would be expected in an examination, others are demonstrated here.

Comparison of Project A with Project C

Equating the £200 000 invested in each of A and C, what return does the extra £300 000 invested in C achieve?

If C is as good as, or better than A and B combined, the extra £300 000 invested in C must achieve a return that matches or exceeds the return from B, that is 14.629 per cent.

Calculation of NPV of incremental cash flow C–A at a discount rate of 14.629 per cent

Year	Inflow A £	Inflow C £	Incremental inflow C–A £	PV of £ at 14.629 per cent	PV of incremental inflow C–A £
0	21 400	40 000	+18 600	1	18 600
1	21 400	40 000	+18 600	0.872	16 219
2	21 400	44 000	+22 600.	0.761	17 199
3	21 400	44 000	+22 600	0.664	15 006
4	21 400	44 000	+22 600	0.579	13 085
5	200 000	640 000	+440 000	0.505	222 200

Present value of incremental inflow C–A	302 309
less Present value of incremental outflow C–A	300 000
NPV of incremental cash flow C–A at 14.629 per cent	+ 2 309

As the NPV is positive, the extra £300 000 invested in C does achieve more than the 14.629 per cent return obtained by investing £300 000 in Project B.

The client is therefore better advised to invest in Project C alone rather than Projects A and B together.

Comparison of Project B with Project C

Equating the £300 000 invested in each of Projects B and C, what return does the extra £200 000 invested in Project C achieve? If Project C is as good as, or better than Projects A and B combined, the extra £200 000 invested in Project C must achieve a return that matches or exceeds the return from Project A, that is 11.98 per cent.

Calculation of NPV of incremental cash flow C-B at a discount rate of 11.98 per cent

Year	Inflow B £	Inflow C £	Incremental inflow C-B £	PV of £1 at 11.98 per cent	PV of incremental inflow C-B
0		40 000	+40 000	1	40 000
1	20 000	40 000	+20 000	0.893	17 860
2	21 600	44 000	+22 400	0.797	17 853
3	23 328	44 000	+20 672	0.712	14 718
4	25 194	44 000	+18 806	0.636	11 961
5	467 210	640 000	+172 790	0.568	98 145
		Present value of incremental inflow C-B			200 537
	less	Present value of incremental outflow C-B			200 000
		NPV of incremental cash flow C-B at 11.98 per cent			+ 537

As the NPV is positive, the extra £200 000 invested in Project C does achieve more than the 11.98 per cent return obtained by investing £200 000 in Project A.

This confirms the conclusion from the previous calculation, that the client would be better advised to invest the whole £500 000 in Project C rather than in Projects A and B together.

Incremental cash flow of Projects A and B combined compared with cash flow of Project C

Year	A + B Inflow £	Outflow £	C Inflow £	Outflow £	Incremental cash flow A + B − C £	PV of £1 at 13.824 per cent	PV A + B − C £
0	21 400	500 000	40 000	500 000	−18 600	1	−18 600
1	41 400		40 000		+ 1 400	0.879	+ 1 231
2	43 000		44 000		− 1 000	0.772	− 772
3	44 728		44 000		+ 728	0.678	+ 494
4	46 594		44 000		+ 2 594	0.596	+ 1 546
5	667 210		640 000		+27 210	0.523	+14 231
	NPV of incremental cash flow A + B − C at 13.824 per cent						− 1 870

The negative NPV indicates that the incremental cash flow produced by combined investment in Projects A and B does not produce the 13.824 per cent obtained from investment solely in Project C. The actual return on the incremental cash flow is 11.304 per cent, calculated as follows:

| | A + B | | C | | Incremental cash flow | PV of £1 at | PV |
Year	Inflow £	Outflow £	Inflow £	Outflow £	A + B – C £	11 per cent	A + B – C £
0	21 400	500 000	40 000	500 000	−18 600	1	−18 600
1	41 400		40 000		+ 1 400	0.901	+ 1 261
2	43 000		44 000		− 1 000	0.812	− 812
3	44 728		44 000		+ 728	0.731	+ 532
4	46 594		44 000		+ 2 594	0.659	+ 1 709
5	667 210		640 000		+27 210	0.593	+16 136

NPV of incremental cash flow A + B – C at 11 per cent   +   226

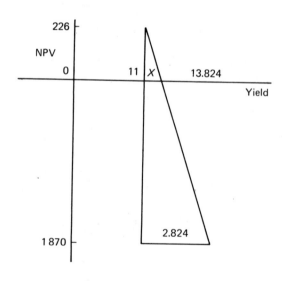

$$\frac{X}{226} = \frac{2.824}{2\,096}$$

$$2\,096X = 638.224$$

$$X = 0.304$$

Incremental IRR of cash flow A + B – C = 11 per cent + 0.304 per cent

= 11.304 per cent.

The incremental yield produced by combined investment in Projects A and B does not achieve the yield of 13.824 per cent from Project C, therefore, once more, the conclusion is that investment in Project C alone is preferred.

An alternative to the above would be to compare the NPV of the income from Projects A and B combined with the IRR of Project C. This is possible since the outlay of Project A together with Project B is exactly the same as that of Project C.

Year	Inflow A + B £	Outflow A + B £	PV of £1 13.824 per cent	PV of inflows £	PV of outflows £
0	21 400	500 000	1	21 400	500 000
1	41 400		0.879	36 391	
2	43 000		0.772	33 196	
3	44 728		0.678	30 326	
4	46 594		0.596	27 770	
5	667 210		0.523	348 951	

	Present value of inflows	498 034
less	Present value of outflows	500 000
	NPV at 13.824 per cent	− 1 966

Again it is confirmed that Project C out-performs Projects A and B together and the client should be advised to invest in Project C alone, in preference to Projects A and B.

Question 2.7

Yield to redemption is the rate of interest which discounts interest and capital payments to equate to market price. In other words, it is the IRR of the investment.

The first step is to calculate the repayments on the original loan of £80 000. It is assumed, in the absence of further information in the question, that this loan covered the whole of the original purchase price.

Calculation of the loan repayments is achieved by dividing the £80 000 by the Years' Purchase single rate at 12 per cent for 15 years.

$$\frac{£80\,000}{YP\ 15\ years\ at\ 12\ per\ cent} = \frac{£80\,000}{6.811} = £11\,746\ per\ annum$$

The amount at present outstanding on the loan is also calculated using the Years' Purchase table.

$$£11\,746 \times YP\ 8\ years\ at\ 12\ per\ cent$$

$$= £11\,746 \times 4.968 = £58\,354$$

Thus, if the investor decides to sell the existing investment and invest the equity elsewhere, he will have to forego

$$£175\,000 - £58\,354 = £116\,646$$

£116 646 must therefore be the amount that he is now investing, whatever course of action he decides to take.

The investor should additionally consider what the present investment may be worth in 8 years' time. The freehold property has increased in value from £80 000 to £175 000 in 7 years. It has therefore increased £175 000/£80 000 = 2.1875 times.

In other words, £80 000 has grown to £175 000, and £1 will grow to £2.1875. Finding $\sqrt[7]{2.1875}$ will reveal the annual rate of growth.

$$\sqrt[7]{2.1875} = 1.118\,315$$

Thus the annual growth rate is 11.8315 per cent.

If this annual growth is maintained, the capital value of the current investment in 8 years' time, will be:

$$£175\,000 \times (1.118\,315)^8$$
$$= £175\,000 \times 2.446$$
$$= £428\,050$$
$$\text{say} \quad £428\,000$$

All the information is now available to calculate the yield to redemption from the existing investment. Some attempt might be made to reflect rental growth in the income flow, but this is difficult since the income is derived from several tenants on different leases, the terms and review dates not being revealed in the question. However, it is obvious when rental increases occur and it could be assumed that the increases shown will be compounded at the rate of growth achieved so far. This possible scenario will first of all be explored.

At the end of year 2, the income rises by £2 000, but it will be assumed that the increase will actually be:

$$£2\,000 \times (1.118\,315)^2 = £2\,501$$

The increase of £4 000 at the end of year 4 would, on the same assumption, be:

$$£4\,000 \times (1.118\,315)^4 = £6\,256$$

The increase at the end of year 6:

$$£2\,000 \times (1.118\,315)^6 = £3\,912$$

The final increase at the end of year 7:

$$£2\,000 \times (1.118\,315)^7 = £4\,375$$

This would produce an income flow over the next 8 years of:

Year	1	2	3	4	5	6	7	8
Income	£12 000	£14 501	£14 501	£20 757	£20 757	£24 669	£24 669	£29 044

This is merely a suggested assumption and is not the only one that might be made.

Calculation of NPV at a discount rate of 20 per cent

Year	Inflow £	Outflow £	PV of £1 at 20 per cent	PV of inflows £	PV of outflows £
0		116 646	1		116 646
1	12 000	11 746	0.833	9 996	9 784
2	14 501	11 746	0.694	10 064	8 152
3	14 501	11 746	0.579	8 396	6 801
4	20 757	11 746	0.482	10 005	5 662
5	20 757	11 746	0.402	8 344	4 722
6	24 669	11 746	0.335	8 264	3 935
7	24 669	11 746	0.279	6 883	3 277
8	29 044 } 428 000 }	11 746	0.233	106 491	2 737
	Present value of inflows			168 443	<u>161 716</u>
	less Present value of outflows			161 716	
	NPV at 20 per cent			+ 6 727	

The yield to redemption must be more than 20 per cent and a rate of 21 per cent will be tested.

Year	Inflow £	Outflow £	PV of £1 at 21 per cent	PV of inflows £	PV of outflows £
0		116 646	1		116 646
1	12 000	11 746	0.826	9 912	9 702
2	14 501	11 746	0.683	9 904	8 023
3	14 501	11 746	0.564	8 179	6 625
4	20 757	11 746	0.467	9 694	5 485
5	20 757	11 746	0.386	8 012	4 534
6	24 669	11 746	0.319	7 869	3 747
7	24 669	11 746	0.263	6 488	3 089
8	29 044 } 428 000 }	11 746	0.218	99 636	2 561
	Present value of inflows			159 694	<u>160 412</u>
	less Present value of outflows			160 412	
	NPV at 21 per cent			− 718	

The yield to redemption must lie between 20 per cent and 21 per cent.

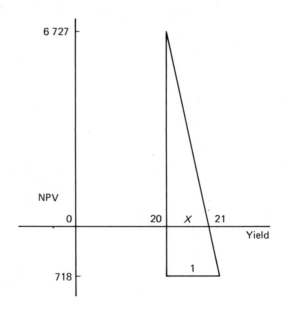

$$\frac{X}{6\,727} = \frac{1}{7\,445}$$

$$7\,445X = 6\,727$$

$$X = 0.904$$

Yield to redemption = 20 per cent + 0.904 per cent

= 20.904 per cent.

The rental flow used in this calculation is based on a forecast using assumptions which may prove to be either an under or overestimation. The investor may consequently have unrealistic expectations of his current investment. It could be argued that the question provides insufficient information upon which to make assumptions regarding future rental growth and that the yield to redemption should be calculated using only the estimated income in current terms. This will produce the worst possible scenario for the investment and arguably a sounder basis for the investor's decision.

Calaculation of NPV at a discount rate of 20 per cent

Year	Inflow £	Outflow £	PV of £1 at 20 per cent	PV of inflows £	PV of outflows £
0		116 646	1		116 646
1	12 000	11 746	0.833	9 996	9 784
2	14 000	11 746	0.694	9 716	8 152
3	14 000	11 746	0.579	8 106	6 801
4	18 000	11 746	0.482	8 676	5 662
5	18 000	11 746	0.402	7 236	4 722
6	20 000	11 746	0.335	6 700	3 935
7	20 000	11 746	0.279	5 580	3 277
8	22 000 〕 428 000 〕	11 746	0.233	104 850	2 737
	Present value of inflows			160 860	161 716
	less Present value of outflows			161 716	
	NPV at 20 per cent			− 856	

The investment does not achieve a return of 20 per cent and a discount rate of 19 per cent will be tested.

Calculation of NPV at a discount rate of 19 per cent

Year	Inflow £	Outflow £	PV of £1 at 19 per cent	PV of inflows £	PV of outflows £
0		116 646	1		116 646
1	12 000	11 746	0.840	10 080	9 867
2	14 000	11 746	0.706	9 884	8 293
3	14 000	11 746	0.593	8 302	6 965
4	18 000	11 746	0.499	8 982	5 861
5	18 000	11 746	0.419	7 542	4 922
6	20 000	11 746	0.352	7 040	4 135
7	20 000	11 746	0.296	5 920	3 477
8	22 000 〕 428 000 〕	11 746	0.249	112 050	2 925
	Present value of inflows			169 800	163 091
	less Present value of outflows			163 091	
	NPV at 19 per cent			+ 6 709	

The yield to redemption must therefore lie between 19 per cent and 20 per cent.

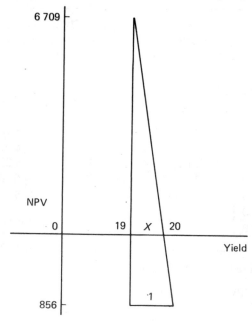

$$\frac{X}{6\,709} = \frac{1}{7\,565}$$

$$7\,565X = 6\,709$$

$$X = 0.887$$

Yield to redemption = 19 per cent + 0.887 per cent

= 19.887 per cent.

   Ignoring any possibilities of potential growth in rental income, the yield to redemption is nevertheless 19.887 per cent and, unless the investor can obtain a return that at least matches this, by investing elsewhere, it appears advisable to retain the equity in the existing freehold property investment.

3 Valuation of Development Properties

The residual method of valuation may be initially adopted when valuing development properties, even though its shortcomings are well established. Slight variations in the different elements in the calculation might produce considerable variations in the final result. Because of this, simple sensitivity testing might be used, as illustrated in Question **3.6.**

More refined techniques such as discounted cash flows, the calculation of Net Present Value and Internal Rate of Return are extensively used, and these give the opportunity for costs and returns to be realistically distributed at regular intervals over a period of time. A discounted cash flow is versatile enough to be used to calculate land value as in Question **3.1,** rents, building costs or developer's profit. The estimation of Internal Rate of Return is an additional step in the use of the discounted cash flow as illustrated in Question **3.2.** This is a useful technique for appraising different development opportunities purely on financial grounds.

It must not be overlooked that considerable amounts of money may be expended on refurbishment, which may be a viable alternative to demolition and redevelopment. Question **3.4** deals with this situation.

One feature of the development process is the creating of acceptable partnership arrangements primarily between local authorities and financial institutions, with the possible inclusion of a third party, the developer. The local authority will own the valuable asset, the land, and the financial institution the funds to create the development. There are various types of agreement, to provide for the sharing of the equity such as 'horizontal' and 'vertical' slicing.

The sale and leaseback transaction is another valuable type of transaction for funding and sharing arrangements.

These different sharing and funding arrangements are illustrated in Question **3.5.**

Computers are a valuable assistance in development appraisal, and spreadsheets are a cheap and versatile tool to be used, as explained in Question **3.6.**

VALUATION OF DEVELOPMENT PROPERTIES – QUESTIONS

3.1. Your client is negotiating to purchase a freehold site, which has detailed planning permission for retail development.

He considers that he can build a shopping centre on the site to provide 50 000 m² gross floor area of retail units comprising three major stores and forty shops.

The following information has been established:

(i) Contract period 36 months

(ii) Estimated cost of construction £500 per m²

(iii) Contingency allowance 10 per cent

(iv) Quantity surveyor's and architect's fees 10 per cent of cost and contingencies (50 per cent to be paid at commencement, 25 per cent on completion and 25 per cent 6 months after completion)

(v) Retention 5 per cent (2½ per cent to be released on completion and 2½ per cent 6 months after completion)

(vi) Financing 3 per cent per quarter

(vii) Anticipated net rents on 5 year leases — £85 per m² of net lettable space

(viii) Anticipated yield — 5½ per cent

(ix) Estate agency, legal fees and advertising — 3 per cent of gross development value.

By the use of a three-monthly discounted cash flow, advise your client as to the maximum amount he should bid for the purchase of the site, assuming a developer's pre-tax profit of £5m.

Ignore taxation implications.

3.2. A building is to be constructed having six shops on the ground floor each with a net frontage of 7.5 m and a net depth of 20 m. There are four upper floors used as offices with a total net lettable area of 2 160 m².

The following information has been established:

(i) Estimated building cost £2m

(ii) Developer's contingencies — 10 per cent of building cost

(iii) Quantity Surveyor's and Architect's fees — 10 per cent of building cost and contingencies (50 per cent to be paid at the commencement and 50 per cent on completion of the building period)

(iv) Retention — 5 per cent (2½ per cent to be released on completion and 2½ per cent 6 months after completion)

(v) Building Period 2 years

(vi) Distribution of building costs:

End of month	per cent
3	10
6	10
9	12
12	12
15	15
18	15
21	14
24	12

(vii) Site purchase and costs £256 000

(viii) Cost of financing 3 per cent per quarter

 (ix) Anticipated rents — Shops Zone A (5 m depth) £250 m^2

                                  Offices £60 m^2

  (x) Anticipated yield 7½ per cent

 (xi) Estate agency, legal fees etc. — 3 per cent of Gross Development Value.

You are required to calculate:

(i)   the net present value of the project using a discounted cash flow technique

(ii)   the internal rate of return for the project.

3.3. (a) The construction of a commercial development with total costs of £470 000 (including land at £50 000 but excluding finance charges) is programmed over 18 months requiring £50 000 at the beginning of the first quarter followed by £70 000 at the end of each quarter.

The construction work has now reached the end of the fourth quarter as programmed. A delay in delivery of materials will extend the construction period by one quarter with estimated costs as follows:

End of fifth quarter	£10 000
End of sixth quarter	£80 000
End of seventh quarter	£70 000

The estimated value on completion is £575 000. Cost of financing is 4 per cent per quarter.

Using appropriate calculations, examine the effect of this delay on the developer's profit.

(b) Various systems of central heating are being considered for inclusion in a major refurbishment.

Calculate the costs-in-use of an oil fired system having an initial cost of £18 000 and running costs of £2 000 per annum. Components costing £6 000 will require replacement at 10 year intervals and components costing £12 000 will require replacement in 20 years' time. The system can be taken as having no value in 40 years' time.

3.4. A polytechnic plans to provide additional student accommodation to be available in 2 years' time. A new building will cost £220 000, will take 2 years to build and will have annual maintenance costs of £2 000 for the first 25 years and £5 000 thereafter.

An alternative is to purchase and refurbish a terrace of old houses at a cost of £180 000. The conversion work will take one year to complete and the annual maintenance costs will be £5 000.

The new building will have a life of 50 years. The conversion will have a life of 25 years but it may be assumed in year 26 similar old houses for conversion could be found and the cost involved in the second generation of old houses is exactly equal to that of the first generation of conversions. You may assume that neither alternative will have any appreciable residual value beyond 50 years.

The £220 000 cost of new building may be assumed to be expended as £110 000 in the first year and as £110 000 in the second year.

Using as the criterion the cost of finance at 13 per cent per annum, determine, on purely financial grounds, which alternative is to be preferred.

3.5. (a) Describe the principal forms of equity participation arrangements used in the financing of property, with appropriate calculations.

(b) Explain the operation of a typical sale and leaseback arrangement in property development, illustrating your answer with an example.

3.6. (a) Discuss the aims and effects of simple sensitivity testing on the residual valuation prepared for a development project. Construct examples to illustrate your answer.

(b) Discuss how spreadsheets can assist when appraising the viability of development schemes.

VALUATION OF DEVELOPMENT PROPERTIES – SUGGESTED ANSWERS

Question 3.1
Preliminary calculations

Building costs 50 000 m² × £500	= £25 000 000
evenly distributed over quarterly periods for 36 months [see note 1]	= £ 2 083 333
Contingencies 10 per cent	= £ 2 500 000
distributed as above [see note 2]	= £ 208 333
Retention – 5 per cent of building cost and contingencies	= £ 1 375 000
evenly distributed over quarterly periods for 36 months	= £ 114 583
Repayment of retention – 2½ per cent [see note 3]	= £ 687 500
Quantity surveyor's and architect's fees 10 per cent of costs	
and contingencies [see note 4]	= £ 2 750 000
(5 per cent = £1 375 000, 2½ per cent = £687 500)	
Net lettable space [see note 5] – deduct 15 per cent from gross space = 42 500 m²	

Gross Development Value	
Net rental value = 42 500 m² × £85	= £ 3 612 500
YP in perpetuity at 5½ per cent [see note 6]	= 18.182
	= £65 682 475

Estate agency, legal fees and advertising = 3 per cent of	
Gross Development Value [see note 7]	= £ 1 970 474
Developer's pre-tax profit [see note 8]	= £5m

Notes

1: For the purpose of this calculation, it is assumed that building costs are evenly distributed throughout the contract period. In practice there would be an uneven distribution, based on an S-curve.

2: The contingency allowance has been evenly distributed alongside the building costs. Again, in practice, contingencies might be required at quite indefinite periods.

3: Within the terms of the building contract, it is usual for a retention to be deducted from the gross value of the work — in this example, 5 per cent. Half the retention is released when the work is completed and the balance 6 months later, when any defects may be required to be remedied.

4: Quantity surveyor's and architect's fees are 10 per cent of building costs and contingencies. It is assumed that, if the contingency allowance is used, fees would be obtainable on this amount.

5: The gross area of 50 000 m² has been reduced by 15 per cent to take account of external and internal walls and circulation areas. The percentage deduction is arbitrary, dependent upon the type of construction.

6: It is assumed that the capital value (Gross Development Value) of the project will be realised after 36 months. Some of the units, however, might not be let immediately, although, in this type of development, there ought to be a substantial number of pre-let units.

7: In this calculation, the fees will be deducted at the 36 month period. In practice, however, some fees may be paid in advance of this.

8: The developer's profit will be inserted in the cash flow on the completion of the project, i.e. at 36 months.

| | Immediate | End of month | | | | |
		3	6	9	12	15
Building costs		-2 083 333	-2 083 333	-2 083 333	-2 083 333	-2 083 333
Contingencies		-208 333	-208 333	-208 333	-208 333	-208 333
Retention		+114 583	+114 583	+114 583	+114 583	+114 583
Release of retention						
QS's and architect's fees	-1 375 000					
Gross development value						
Estate agency's etc. fees						
Developer's profit						
Total	-1 375 000	-2 177 083	-2 177 083	-2 177 083	-2 177 083	-2 177 083
Discounted at 3 per cent per quarter	1.0	0.971	0.943	0.915	0.888	0.863
Total flow	-1 375 000	-2 113 947	-2 052 989	-1 992 030	-1 933 249	-1 878 822

18	21	24	27	30	33	36	42
−2 083 333	−2 083 333	−2 083 333	−2 083 333	−2 083 333	−2 083 333	−2 083 333	
−208 333	−208 333	−208 333	−208 333	−208 333	−208 333	−208 333	
+114 583	+114 583	+114 583	+114 583	+114 583	+114 583	+114 583	
						−687 500	−687 500
						−687 500	−687 500
						+65 682 475	
						−1 970 474	
						−5 000 000	
−2 177 083	−2 177 083	−2 177 083	−2 177 083	−2 177 083	−2 177 083	+55 159 918	−1 375 000
0.837	0.813	0.789	0.766	0.744	0.722	0.701	0.661
−1 822 218	−1 769 968	−1 717 718	−1 667 645	−1 619 750	−1 572 374	+38 667 100	−908 875

$$\text{Total flow} = + £38\,667\,100 - £22\,424\,585$$

$$= + £16\,242\,515$$

This represents the amount available for land purchase and acquisition costs say 4 per cent.

$$\text{So site value} = \frac{£16\,242\,515}{1.04} = £15\,617\,802$$

say £15 617 800

Question 3.2

(i) *Preliminary calculations*

Building cost £2m + contingencies (10 per cent of £2m) = £2 200 000
Retention 5 per cent of £2 200 000   = £   110 000

It is assumed that building costs, contingencies and retention are all distributed over the 2 year building period as stated in the question:

Period months	Percentage	Building cost + contingencies	Retention	Balance
3	10	£220 000	£11 000	£209 000
6	10	£220 000	£11 000	£209 000
9	12	£264 000	£13 200	£250 800
12	12	£264 000	£13 200	£250 800
15	15	£330 000	£16 500	£313 500
18	15	£330 000	£16 500	£313 500
21	14	£308 000	£15 400	£292 600
24	12	£264 000	£13 200	£250 800
	100	£2 200 000	£110 000	£2 090 000

Quantity Surveyor's and Architect's Fees
  10 per cent of £2 200 000 = £220 000

Calculation of Gross Development Value —
Shops using two 5 m zones and a remainder and 'halving back':

7.5 × 5 × £250	=	£9 375
7.5 × 5 × £125	=	£4 687.5
7.5 × 10 × £62.5	=	£4 687.5
Rental value of one shop	=	£18 750 pa

× 6		
Value of shops	=	£112 500 pa
Offices — 2160 m^2 net × £60	=	£129 600 pa
Full rental value	=	£242 100 pa
YP in perpetuity at 7½ per cent	=	13.333
		£3 227 919
Gross Development Value say		£3 228 000

Estate agency, legal fees etc. - 3 per cent of £3 228 000 = £96 840

The above information may now be incorporated in a three-monthly cash flow to obtain the Net Present Value.

Item	Immediate				End of month					
		3	6	9	12	15	18	21	24	30
Site purchase and costs	−256 000									
Building cost and contingencies *less* retention		−209 000	−209 000	−250 800	−250 800	−313 500	−313 500	−292 600	−250 800	
QS's and architect's fees	−110 000								−110 000	
Release of retention									−55 000	−55 000
Gross development value									+3 228 000	
Estate agency and legal fees									−96 840	
Total	−366 000	−209 000	−209 000	−250 800	−250 800	−313 500	−313 500	−292 600	+2 715 360	−55 000
Discounted at 3 per cent per quarter	1.0	0.971	0.943	0.915	0.888	0.863	0.837	0.813	0.789	0.744
Total flow	−366 000	−202 939	−197 087	−229 482	−222 710	−270 550	−262 399	−237 884	+2 142 419	−40 920

Total Flow = £2 142 419 – £2 029 971

= +£112 448

Net Present Value = £+112 448

(ii) The Internal Rate of Return (IRR) is the rate of discount, which will reduce the present value of cash flows to the same value as the cash invested in the project, i.e. a Net Present Value of zero.

To calculate the IRR, assume a discount rate higher than 3 per cent per quarter, say 5 per cent per quarter, and discount the flow at this rate:

End of month	Flow	Discount at 5 per cent per quarter	Discounted flow
Immediate	−366 000	1.0	−366 000
3	−209 000	0.952	−198 968
6	−209 000	0.907	−189 563
9	−250 800	0.864	−216 691
12	−250 800	0.823	−206 408
15	−313 500	0.784	−245 784
18	−313 500	0.746	−233 871
21	−292 600	0.711	−208 039
24	+2 715 360	0.677	+1 838 298
30	−55 000	0.614	−33 770
			+1 838 298
			−1 899 094
		Net Present Value =	−£60 796

The IRR lies between 3 per cent and 5 per cent and may be found by similar triangles.

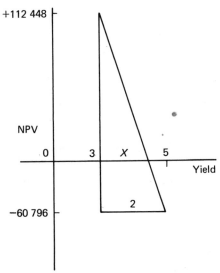

$$\frac{X}{112\,448} = \frac{2}{173\,244}$$

$$X = \frac{2 \times 112\,448}{173\,244}$$

$$= 1.298$$

Internal Rate of Return $= 3 + 1.298$

$= 4.298$ per cent

Question 3.3

(a) This requires the calculation of developer's profit in accordance with the original scenario.

This may be achieved by taking each cost plus its interest at 4 per cent per quarter, and deducting the total from the value of £575 000

i.e. Value in 18 months   £575 000

Costs

£50 000 × Amount of £1 — 6 periods
at 4 per cent = £50 000 × 1.265   = £63 250

£70 000 × Amount of £1 — 5 periods
at 4 per cent = £70 000 × 1.217   = £85 190

£70 000 × Amount of £1 — 4 periods
at 4 per cent = £70 000 × 1.17   = £81 900

£70 000 × Amount of £1 — 3 periods
at 4 per cent = £70 000 × 1.125   = £78 750

£70 000 × Amount of £1 — 2 periods
at 4 per cent = £70 000 × 1.082   = £75 740

£70 000 × Amount of £1 — 1 period
at 4 per cent = £70 000 × 1.04   = £72 800

Last quarter   = £70 000   £527 630

                    Developer's profit   £ 47 370

An alternative method is to discount the value and costs at 4 per cent per quarter, which will give the Net Present Value. This figure will then have compound interest for 6 periods at 4 per cent per quarter to give the developer's profit.

Inflow

Value in 18 months	£575 000	
PV of £1 6 periods at 4 per cent	0.79	£454 250

Outflow

Quarterly cost	£ 70 000	
YP 6 years at 4 per cent [see note 1]	5.242	
	£366 940	
Initial cost	£ 50 000	£416 940
Net Present Value		= £ 37 310
Amount of £1 − 6 periods at 4 per cent		= 1.265
Developer's profit [see note 2]		£ 47 200

Notes

1: An alternative would be to discount each £70 000 with the PV of £1 for each quarter at 4 per cent. The single rate YP is the summation of each PV.

2: The difference of £170 compared with the first method is due to 'rounding-off'.

The calculation should now be repeated in accordance with the revised scenario

i.e. Value in 21 months [see note 1]	£575 000

Costs

£50 000 × Amount of £1 − 7 periods at 4 per cent = £50 000 × 1.316	= £65 800
£70 000 × Amount of £1 − 6 periods at 4 per cent = £70 000 × 1.265	= £88 550
£70 000 × Amount of £1 − 5 periods at 4 per cent = £70 000 × 1.217	= £85 190
£70 000 × Amount of £1 − 4 periods at 4 per cent = £70 000 × 1.17	= £81 900
£70 000 × Amount of £1 − 3 periods at 4 per cent = £70 000 × 1.125	= £78 750
£10 000 × Amount of £1 − 2 periods at 4 per cent = £10 000 × 1.082	= £10 820
£80 000 × Amount of £1 − 1 period at 4 per cent = £80 000 × 1.04	= £83 200
Last quarter	= £70 000
	£564 210

Developer's profit	£10 790

Note

1: It is assumed that the value of £575 000 will still be the same in 21 months.

The delay in deliveries will reduce the developer's profit from £47 370 to £10 790, and the profit will not be realised until 3 months later.

(b) The calculation of costs-in-use of an oil fired central heating system requires that all costs be converted to an annual basis; in many cases, annual equivalents will be obtained.

Assuming an interest rate of 5 per cent

Initial cost	£18 000

Component (i) — £6 000 × (PV of £1 in 10 years at 5 per cent + PV of £1 in 20 years at 5 per cent + PV of £1 in 30 years at 5 per cent) = £6 000 × (0.614 + 0.377 + 0.231) = £6 000 × 1.222   = £ 7 332

Component (ii) — £12 000 × PV of £1 in 20 years
at 5 per cent = £12 000 × 0.377   = £ 4 524

    Present day liability   = £29 856

$$\text{Annual equivalent} = \frac{£29\,856}{\text{YP 40 years at 5 per cent}}$$

$$= \frac{£29\,856}{17.159} \qquad = £\ 1\,740\ \text{pa}$$

Running costs   = £ 2 000 pa

    Costs-in-use   = £ 3 740 pa

Note

The purpose of calculating the costs-in-use of one type of central heating is that it may be compared with the costs-in-use of other heating types. A decision may be made as to the type of system to be installed according to the most economic cost.

While the technique is mathematically acceptable, it is based on present-day circumstances and cannot take account of future fluctuations in cost. There is also considerable skill needed in estimating the life of the asset — in this case, 40 years.

It is usual to calculate the annual equivalents on a Years' Purchase single rate basis. Because of the limited life of the asset, a case might be made for the use of a dual rate tax-adjusted Years' Purchase.

Question 3.4

This question requires a calculation, where each alternative has its costs discounted at 13 per cent per annum to obtain the net present cost, the lesser figure giving the better alternative.

It is beyond the requirements of the question to compare the schemes in relation to such factors as appearance, quality and efficiency, as this would form part of a cost-benefit analysis exercise.

Scheme 1 — New building

	Cost per annum	Discounted cost
Year 1	£110 000 × PV in 1 year at 13 per cent (0.885)	= £ 97 350
Year 2	£110 000 × PV in 2 years at 13 per cent (0.783)	= £ 86 130
Year 3 to Year 25	£2 000 × YP 23 years at 13 per cent × PV in 2 years at 13 per cent [see note 1] (7.23 × 0.783)	= £ 11 322
Year 26 to Year 50	£5 000 × YP 25 years at 13 per cent × PV in 25 years at 13 per cent [see note 2] (7.33 × 0.047)	= £ 1 723
	Discounted cost	£196 525

Notes

1: This calculation will give the same result as discounting each £2 000 with PV of £1 for each year from 2 years to 25 years.
2: This will give the same result as discounting each £5 000 with PV of £1 for each year from 25 years to 50 years.

Scheme 2 — Conversion

	Cost per annum	Discounted cost
Year 1	£180 000 × PV in 1 year at 13 per cent (0.885)	= £159 300
Year 2 to Year 25	£5 000 × YP 24 years at 13 per cent × PV in 1 year at 13 per cent [see note 1] (7.283 × 0.885)	= £ 32 227
Year 26	£180 000 × PV in 26 years at 13 per cent (0.042)	= £ 7 560
Year 27 to Year 50	£5 000 × YP 24 years at 13 per cent × PV in 26 years at 13 per cent [see note 2] (7.283 × 0.042)	= £ 1 529
	Discounted cost	= £200 616

Notes

1: This calculation will give the same result as discounting each £5 000 with PV of £1 for each year from 1 year to 25 years.

2: This will give the same result as discounting each £5 000 with PV of £1 for each year from 26 years to 50 years.

The new building gives a lower cost (£196 525 compared with £200 616) than the refurbishment so that this is recommended as the better scheme.

Question 3.5

(a) The first step when answering this question may be to construct a scenario.

For example, a local authority owns the freehold of land which is suitable for commercial development, but it has insufficient finances to develop the site itself. It would, therefore, welcome partnership with an institution, who would take a lease of the site but fund the scheme. Assume the estimated net rental value of the completed development is considered to be £750 000 per annum and the costs are estimated at £5m. The site would be let on a 125 year lease, with 5 year rent reviews.

The institution, as lessee, will expect from lettings to initially receive £750 000 per annum, but they will pay a ground rent per annum to the local authority, taking their funding commitment into account.

i.e. Estimated full rental value	£750 000 pa
Institutional return – 8 per cent on £3m	
[see note 1]	£240 000 pa
Amount available for ground rent per annum	£510 000 pa

Note

1: Initial yield is considered to be 6 per cent with an addition of 2 per cent to reflect risk and management.

The initial equity share as a percentage is

$$\frac{\text{Ground rent}}{\text{Current rental value}} \times 100$$

$$= \frac{510\,000}{750\,000} \times 100 = 68 \text{ per cent}$$

Future ground rent reviews may be 'geared' to the same percentage, i.e. ground rent will be 68 per cent of net rental value when reviewed in the future.

The eventual income and costs of development on completion of the scheme may be different from those originally calculated.

To avoid 'missing out' on increased rental value a participation clause may be incorporated in the partnership agreement, whereby the local authority and institution share in increased rental income at an agreed percentage.

Assume in the previous example, that, on completion, rents are £900 000 per annum and costs £4m. The parties agree to equal sharing.

then: Rents on completion	= £900 000 pa
8 per cent on £4m	= £320 000 pa
Actual profit rent	= £580 000 pa
Less Agreed minimum ground rent	= £510 000 pa
Amount available for sharing	£ 70 000 pa
Local authority share at 50 per cent	= £ 35 000 pa
Plus Initial ground rent	= £510 000 pa
Revised ground rent	£545 000 pa

If this is then used for gearing the percentage will be

$$\frac{£545\,000}{£900\,000} \times 100 = 60.5 \text{ per cent}$$

Thus, the equity share has moved slightly towards the institution.

The local authority may also protect itself in the eventuality of the actual rent/cost ratio being unfavourable. It will have a 'yield protection clause' which guarantees the original ground rent.

For example, if the rents received total £800 000 but costs are £4.5m, the local authority will still require a ground rent of £510 000 per annum:

Rents on completion	£800 000 pa
8 per cent on £4.5m	£360 000 pa
	£440 000 pa
Less Agreed ground rent	£510 000 pa
Deficit	−£ 70 000 pa

Revised equity share to local authority would be

$$\frac{£440\,000}{£800\,000} \times 100 = 55 \text{ per cent}$$

The different types of sharing arrangements may be illustrated as follows:

(A)

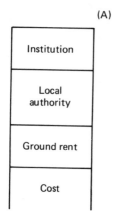

Horizontal sharing
In this arrangement, the local authority receives a fixed initial amount of the equity, leaving the institution ungeared with the riskier 'top slice'.

If, however, there is greater growth than anticipated, the 'top slice' share (the institution) will benefit.

(B)

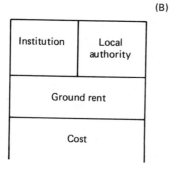

Vertical sharing
In this arrangement, the equity is shared 'side-by-side' in pre-determined proportions. The local authority participates in possible income growth, but equally has risks along with the institution in a low growth situation.

Third party inclusion

A financial arrangement may include a third party — the developer or development company. This may be undertaken on a 'fixed horizontal basis' or a 'side-by-side basis'.

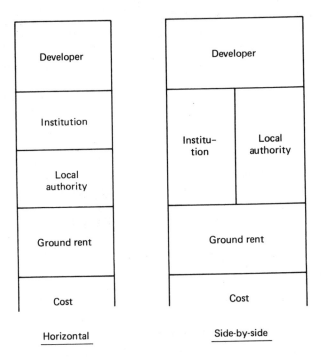

Horizontal   Side-by-side

In both cases, the motives of the parties are the same.

The local authority may establish planning policies, prepare development guide-lines, assist in land assembly and provide the infrastructure.

In return, it will expect a well-secured bottom slice ground rent reviewed at regular intervals plus equity participation.

The institution's concern is the supplying of long-term investment funds and arranging short-term development finance. It receives a medium slice long leasehold interest after the local authority (horizontal sharing) or alongside the local authority (side-by-side sharing).

The development company allows for a normal profit margin within its development costs and may accept a top slice which rewards for identifying and initiating the scheme, project supervision, marketing and management.

Sometimes, a fee for project supervision is levied ahead of the ground rent or as a prior charge upon the equity.

(b) Sale and leaseback might be the most important financial arrangement for sharing the equity of development projects, as it is a method of raising significant capital plus a share in the revenue of the development.

The developer sells the freehold to an institution and takes back a long lease, with the developer subletting the property to occupying tenants and obtaining a profit rent. Rent reviews may be seven, five or three years.

The developer obtains total permanent finance, immediate profit rent and continuing participation in the project. He has no need to find any bridging finance, and he has an interest capable of sale, from which the proceeds could finance other projects. Disadvantages might be a complicated rental structure and tax consequences may be difficult.

The arrangement gives the investor substantial equity involvement in newly-built property with a secure tenant, who will be responsible for the collection of rents from sublessees and who will be responsible for the bulk of management costs.

An example will illustrate this:

Assume an institution buys the freehold site from a developer for £¾m, and leases back the completed development, which has an anticipated net rental value of £100 000 per annum. The institution requires a return of 7 per cent on capital expended, plus 50 per cent of all rental income achieved in excess of £100 000.

If the rent achieved is £120 000 per annum initially, then the ground rent per annum will be:

Leaseback initial rent 7½ per cent of £¾m	= £56 250 pa
plus Excess rent 50 per cent of £20 000	£10 000 pa
Initial ground rent	= £66 250 pa

The letting is usually on a long lease basis which will be subject to rent reviews, so that the division of rental income is significant.

If an arrangement is made where the developer receives 'top slice income', this is available after a minimum guarantee of income to the institution. If the income falls below the guaranteed return to the institution, the developer will sacrifice the top slice.

A side-by-side sharing arrangement may be more beneficial to a developer as the rental income is shared in agreed proportions.

It may be possible to incorporate a gearing arrangement, which may give advantage to the developer in the sale and leaseback arrangement.

Referring to the above example, the institution receives £66 250 per annum (i.e. 55.2 per cent) and the developer a profit rent of £53 750 per annum (44.8 per cent). If these proportions are applied to future rent reviews, both parties participate in growth but the developer would not increase his overall percentage of income from the development he has created. This ungeared arrangement may be illustrated below:

Imagine rental income per annum when subsequently reviewed is £150 000 per annum, then the institution would receive 55.2 per cent of this, i.e. £82 800 per annum and the developer would receive 44.8 per cent, i.e. £67 200 per annum.

If the arrangement is left on the original geared arrangement, the calculation is:

Leaseback initial rent 7½ per cent of £¾m	= £56 250 pa
Plus Excess rent 50 per cent of £50 000	= £25 000 pa
Initial ground rent	= £81 250 pa

The developer has a profit rent of £68 750 pa, so he is slightly favoured by the geared arrangement.

A characteristic of the sale and leaseback structure is the speculative nature of estimated future income. The institution might prefer a situation where the property has been pre-let so that there is a guaranteed initial rental income, although much will depend on the institution's confidence in the development.

Question 3.6

(a) A residual valuation is very sensitive to slight variations in its different elements such as rent, initial yield, construction costs, financing rate and building period. Because of this, the Lands Tribunal has, at times, regarded the method as one of 'last resort' (*First Garden City Ltd v. Letchworth Garden City Corporation, 1966*).

The aim of sensitivity testing is to examine the effects of changes in variables on the residual value, and the basic method involves changing one variable at a time, recalculating the valuation and then analysing the result. The percentage change in the variable is compared with the percentage change in the residual amount. If a small percentage change in the variable produces a large percentage change in the residual amount, then this variable is very sensitive. The technique is basic, but it will give the developer extra information and may have some effect on his decision making.

Example

Construct a residual valuation for the provision of 1200 m² net of office space — building period 1 year.

Then 1 200 m² × £80 [see note 1]		=	£96 000 pa
YP in perpetuity at 6 per cent		=	16.667
Gross Development Value	say		£1 600 000

Less

(i) Building costs 1 320 m² gross at £500 per m² [see note 2]	= £660 000	
(ii) Contingencies 10 per cent of building costs	= £ 66 000	
(iii) Quantity surveyor's and architect's fees — 10 per cent of building costs and contingencies	= £ 72 600	
(iv) Interest charges say 12 per cent on half (i), (ii) and (iii) for 1 year [see note 3]	= £ 47 900	
(v) Estate agency and legal fees 3 per cent of Gross Development Value	= £ 48 000	
(vi) Developers Profit 10 per cent of Gross Development Value	= £160 000	£1 054 500
Amount available for land purchase, acquisition costs and finance		£ 545 500

If acquisition costs are 4 per cent, and financing is 12 per cent, then

$$\text{Land value} = \frac{£545\,500}{1.1648\ [\text{see note 4}]} = £468\,320$$

Notes

1: It is assumed that a rental of £80 per m^2 is appropriate for this type of property.
2: The net area of 1200 m^2 has been increased by 10 per cent to give 1320 m^2 gross and a reasonable building cost is £500 per m^2.
3: It is assumed that building costs, contingencies and fees are distributed evenly throughout the year, so that financing is averaged — half the costs required over the year.
4: 1.1648 is the value of the land and costs (1.04) with interest at 12 per cent on it, i.e. 1.04 × 1.12 = 1.1648.

Assume that the yield is lowered by 1 per cent, i.e. a 16.667 per cent change, then the valuation becomes

Rental value as before	£ 96 000	pa
YP in perpetuity at 5 per cent =	20	
Gross Development Value	£1 920 000	

Less

Items (i)–(iv) as before	= £	846 500
(v) Estate agency 3 per cent of £1 920 000	= £	57 600
(vi) Developer's profit 10 per cent of £1 920 000	= £	192 000
		£1 096 100
Residue		£ 823 900

$$\text{Land value} = \frac{823\,900}{1.1648} = £707\,331$$

Difference in yields is 16.667 per cent but difference in site value is

$$\frac{707\,331 - 468\,320}{707\,331} = 33.8 \text{ per cent}$$

so that this variable is very sensitive.

A further illustration might be shown by increasing the building costs by 10 per cent.

So, the calculation becomes:

Gross Development Value as before   £1 600 000

Less
Items (i)–(iv) increased by 10 per cent   = £931 150
Items (v) and (vi) as before   = £208 000

                                                                                                £1 139 150

     Residue   £ 460 850

$$\text{Land value} = £\ \frac{460\,850}{1.1648} = £395\,647$$

Difference in building cost is 10 per cent, but difference in site value is

$$\frac{468\,320 - 395\,647}{468\,320} = 15.5 \text{ per cent}$$

so that this variable is fairly sensitive.

This approach is then applied to the other variables and results analysed. Two variables may be altered at the same time, looking at the effect on the residual answer.

(b) Spreadsheets may be used to produce residual valuations, cashflow forecasts, cashflow valuations and appraisals.

They may be used to carry out:

 (i) valuations for alternative scenarios

 (ii) sensitivity testing — particularly useful for this purpose

(iii) 'S' curve calculations — this is the distribution of building costs during the construction period.

Spreadsheets are cheap, versatile and adaptable, and more sophisticated ones have functions that enable the production of such displays as graphs, histograms and pie charts.

They may not be suitable for simulation exercises, and comprehensive sensitivity testing may require the use of more sophisticated software.

4 Taxation

This section deals with various aspects of taxation involving Capital Gains Tax, Inheritance Tax and Value Added Tax.

Calculation of the chargeable gain for Capital Gains Tax purposes has been complicated rather than simplified by the *Finance Act 1988*, which changed the base date for calculating such gains from 6 April 1965 to 31 March 1982. However, if the chargeable gain produced by using this base date is higher than would have been produced under the provisions existing prior to the *Finance Act 1988*, then the latter is adopted.

Consequently, two calculations are now necessary in order to ascertain the most favourable for the taxpayer. Question **4.1** illustrates this in the situation where a disposal of part of a freehold interest occurs, the whole having been acquired prior to 6 April 1965, and Question **4.2** considers disposal subject to a previous gain that has been rolled over. Leasehold interests are treated by the Inland Revenue as wasting assets when the lease has 50 years or less unexpired, and Question **4.3** demonstrates how the receipt of a premium on sub-letting is dealt with for the purposes of both Capital Gains Tax and Schedule A Income Tax.

Until recent years, tax on a person's estate on death was regarded as largely voluntary, since by careful planning it was possible to mitigate the full impact of the tax. This situation was changed when Capital Transfer Tax replaced Estate Duty in 1975. Lifetime gifts became chargeable as they occurred and the tax was cumulative, so that successive gifts were charged at higher rates of tax, ending with the final disposal on death. In 1986, Capital Transfer Tax was replaced by Inheritance Tax, which largely returned the situation to that which had existed under Estate Duty. With some exceptions, lifetime transfers are potentially exempt, becoming completely exempt if the donor survives for 7 years after the transfer. Otherwise, such transfers of value are charged at the rate of tax applicable at the date of death. The principles applied in the valuation of assets for Inheritance Tax purposes are examined in Question **4.4**.

Finally, consideration is given in this section to Value Added Tax. In June 1988, the European Court of Justice decided that VAT should be charged on various property transactions which had previously been zero rated. The Government published its draft proposals for implementing this on 6 February 1989. The impact is mainly on commercial property, since the present zero rating for construction of dwellings is largely unchanged.

At the time of writing, the effect of VAT on values and valuation is purely conjecture. There are various questions that will hopefully be clarified in due course. For instance, in a valuation, should rents be considered net of VAT or gross, as in the case of Income Tax, adjusting where necessary to reflect the position of individual investors or tenants? Certainly the process of considering comparable information will be more complicated. The valuer will need to ensure that all information is on a similar basis, whether net or gross of VAT and, in future, the type of tenant is likely to be much more crucial. The VAT provisions will also no doubt create difficulties in the rent review process. The two questions provided do not pretend to give answers to all the many problems which may emerge. Question **4.5** considers the likely effects upon a developer and Question **4.6** examines the possible effects upon both the landlord and his various tenants as a consequence of the landlord's decision regarding election to charge VAT upon rents in an existing building.

TAXATION – QUESTIONS

4.1. Mr Robinson acquired a vacant bombed site in April 1949 for £500. In March 1965, he applied for, and was granted, planning permission to use the site for car parking. At that time the property was valued at £10 000.

In July 1966 he spent £5 000 on laying tarmacadam, in August 1967, £1 000 on marking out spaces, in 1969 £300 on repairs to the surface and similar amounts in 1971 and 1973.

In March 1982 the value of the property was £100 000.

Mr Robinson applied for planning permission in February 1986 to erect an office building on half the site and on appeal consent was granted, costs of appeal being £2 500. In July 1987, Mr Robinson spent £6 000 on providing automatic barriers for the car park. In January 1988, development of the office building commenced. The costs of development were £1 200 000 and the completed development was sold at the end of March 1989 for £5 000 000. The value of the remaining site used as a car park was £75 000.

Assuming the Retail Prices Index in March 1982 was 100, in January 1988 was 126.3, in February 1986 was 113.7 and in March 1989 was 132.6, calculate the Capital Gains Tax liability of Mr Robinson.

4.2. Timbo Pet Supplies occupy a leasehold warehouse which they purchased in 1978 for £155 000 when the lease had 70 years unexpired following the compulsory acquisition of their former freehold premises. The former premises had been purchased in 1955 for £17 000 and were improved in 1960 by the addition of a canteen and rest room at a cost of £6 000. The compensation they received in 1978 totalled £78 000, comprising £70 000 for the freehold premises plus £8 000 disturbance costs. This compensation was then applied to the purchase of their present premises.

Timbo Pet Supplies are now considering the transfer of their business to other property which they own and have received an offer of £245 000 for their present leasehold warehouse.

Advise Timbo Pet Supplies of the likely chargeable gain for Capital Gains Tax purposes if they go ahead with the sale at £245 000. You may assume that the value of their interest in March 1982 was £190 000, the Retail Prices Index in March 1982 was 100 and at the date of sale 134.3.

4.3. Byron granted a 50 year lease to Goldsmith in April 1964 at a fixed rent on full repairing and insuring terms of £30 000 per annum, together with an initial premium equal to 1 year's rent.

Improvements to the premises were carried out in September 1980 at a cost of £12 000. In March 1989, Goldsmith granted Sheridan a 20 year lease at a fixed rent of £65 000 per annum on full repairing and insuring terms, together with an initial premium of £100 000.

The value of Goldsmith's interest at 31 March 1982 was £230 000, the Retail Prices Index in March 1982 was 100 and in March 1989 was 132.6.

Making any reasonable assumptions you deem necessary, calculate the chargeable gain for Capital Gains Tax purposes on receipt of the £100 000 by Goldsmith.

4.4. You have to deal with the assessment of a freehold estate for taxation purposes upon the recent death of the owner. The deceased purchased the estate in 1975, and the estate comprises:

A farm of 116 hectares farmed by the owner, who lived in the farmhouse; this farm has development potential.

Two farms of 88 hectares and 93 hectares respectively, each let to a tenant.

Ten houses in a nearby town, each let at £30 per week exclusive, the landlord being responsible for external repairs.

Six shops in the same town each let 6 years ago on a 20 year lease, with 5 year rent reviews on full repairing and insuring terms. In each case, the current rent passing is £26 500 per annum and the full net rental value is £28 000 per annum.

(a) Explain the basis of assessment.
(b) Discuss the principles of valuation.
(c) Produce an outline valuation of the whole estate.

You may make any reasonable assumptions in answering this question.

4.5. A developer is considering the purchase of freehold land for development with shops and offices and has asked your advice regarding the likely profit he will achieve from the scheme:

(i) if he elects to subject rents on letting to VAT
(ii) if he does not so elect.

Details of the proposed development are, as follows:

Value of completed development	£12 500 000
Building costs	£ 3 000 000
Purchase of site	£ 4 500 000
Costs of site purchase	4 per cent
Development period	2 years

Finance is available for the development at 16 per cent per annum.
Advise the developer.

4.6. Freehold property comprising a ground floor shop with two floors of offices over, is let as follows:

	Ground floor shop	First floor offices	Second floor offices
Tenant	Charity	Insurance Company	Firm of Solicitors
Rent passing	£25 000 pa	£23 000 pa	£20 000 pa
Unexpired term of lease	4 years	4 years	4 years
Current rental value	£32 000 pa	£30 000 pa	£27 500 pa

All rents are on full repairing and insuring terms. Consider the effects upon both landlord and tenants if the landlord elects to charge VAT on the rents.

TAXATION – SUGGESTED ANSWERS

Question 4.1

It should first of all be explained that under the provisions of the *Finance Act 1988*, 31 March 1982 became the base date for the calculation of chargeable gains. In other words, only the gain in value after 31 March 1982 is chargeable, with an allowance for inflationary gains since that date. However, if the gain calculated using the provisions prior to the *Finance Act, 1988* is lower, then this lower gain will be adopted instead.

In answering this question it is therefore necessary to produce 3 calculations using:

  (i) value at March 1982
 (ii) time apportionment
(iii) value at April 1965.

Each of these calculations is shown below. Incidental costs are ignored throughout, but in practice would be set against the chargeable gain.

Calculation of chargeable gain using value at March 1982

Sale Price			£5 000 000
less Value in March 1982 [see note 1]	£50 000		
Costs of planning appeal in 1986 [see note 2]	£ 2 500		
Costs of development	£1 200 000		£1 252 500
Gain in value since March 1982			£3 747 500
less Indexation allowance			

$$£50\,000 \times \frac{132.6 - 100}{100} = \qquad £16\,300 \text{ [see note 3]}$$

$$£2\,500 \times \frac{132.6 - 113.7}{113.7} = \qquad £\quad 416 \text{ [see note 4]}$$

$$£1\,200\,000 \times \frac{132.6 - 129.45}{129.45} = \qquad £29\,200 \qquad\qquad £\quad 45\,916$$

[see note 5]

Chargeable gain	£3 701 584

Notes

1: Value of whole site in March 1982 was £100 000. Since the office development is on only half the site, its value in 1982 is taken as half the total. Section 35 of the *Capital Gains Act 1979* allows apportionment of expenditure in a "manner which is just and reasonable."

Any expenditure prior to 31 March 1982 is disregarded, since it will be reflected in the March 1982 value.

2: Planning appeal was only in respect of the land which has been developed and disposed of, therefore all the costs of this item will be set against the gain.

3: Allowance for inflationary gains since 31 March 1982. Introduced by the *Finance Act 1982* and amended by the *Finance Act 1985*. Allowance is calculated using the rise in the Retail Prices Index as a yard stick. Value at March 1982 is multiplied by

$$\frac{RD - RI}{RI}$$

where

RD = Retail prices index for month of disposal.
RI = Retail prices index for March 1982.

4: Expenditure after 31 March, 1982 is indexed from the date expenditure was incurred to date of disposal.

5: Actual dates of expenditure are not known, apart from the fact that it commenced in January 1988 and the development was sold in March 1989. The average RPI for January 1988 and March 1989 has been adopted here, but in practice the valuer would have more precise knowledge of exactly when expenditure occurred.

Calculation of chargeable gain using time apportionment

		Sale price		£5 000 000
less	Purchase price	£	250	
1966	Cost of laying tarmacadam	£	2 500	
1967	Cost of marking out spaces	£	500	[see note 1]
1969, 1971 and				
1973	Cost of repairs to surface	£	450	
1986	Costs of planning appeal	£	2 500	[see note 2]
1988	Costs of development	£1 200 000		£1 206 200
		Gain from 1949		£3 793 800

less Indexation allowance

Value at March 1982	£	50 000
Expenditure up to 31 March 1982	£	3 700
therefore adopt [see note 3]	£	50 000

$$£50\,000 \times \frac{132.6 - 100}{100} = \qquad £ \quad 16\,300$$

$$£2\,500 \times \frac{132.6 - 113.7}{113.7} = \qquad £ \quad 416$$

$$£1\,200\,000 \times \frac{132.6 - 129.45}{129.45} = \qquad £ \quad 29\,200$$

[see note 4]

	£	45 916
Indexed gain from 1949		£3 747 884

Proportion of gain attributable to [see note 5]:

Purchase	$£3\,747\,884 \times \dfrac{£250}{£1\,206\,200} =$	£	777
Expenditure after 6 April 1965	$£3\,747\,884 \times \dfrac{£1\,205\,950}{£1\,206\,200} =$		£3 747 107

Time apportionment [see note 6]

Purchase [see note 7]	$£777 \times \dfrac{24}{16 + 24} =$		£	466
Expenditure after 6 April 1965	Gain all chargeable [see note 8]			£3 747 107
	Chargeable gain			£3 747 573

Notes

1: See note 1 of calculation using value at March 1982. The same principle has been applied to actual expenditure.

2: See note 2 of calculation using value at March 1982.

3: Under provisions prior to *Finance Act 1988*, indexation allowance is applied to the greater of the value at 31 March 1982 or the total expenditure up to that date.

4: See note 5 of calculation using value at March 1982.

5: Calculation to separate portions of total gain that have occurred before and after 6 April 1965. The various items of expenditure have been incurred at different times, therefore gains caused by these expenditures have accrued over varying lengths of time and need to be separated. This separation is achieved mathematically by

$$\text{total gain} \times \frac{\text{individual item of expenditure}}{\text{total expenditure}}$$

In this case, only purchase occurred prior to 6 April 1965, all other items of expenditure occurred from 1966 onwards and can be considered in total.

6: The gain of £3 747 573 has been accruing since April 1949 and any gain achieved prior to 6 April 1965 must be excluded since it is not chargeable. Using the time apportionment method, the capital gain after 6 April 1965 is assumed to have occurred in the same proportion as time after 6 April 1965 compared to the total period of ownership.

7: Time apportionment formula $\dfrac{T}{P + T}$ where

T = time between 6 April 1965 to date of disposal

P = time between date of expenditure and 6 April 1965.

8: Only gains accruing since 6 April 1965 are chargeable. Since this expenditure took place at various times from 1966 onwards, any gain in value resulting from the expenditure must have occurred after 6 April 1965 and is therefore all chargeable.

Calculation of chargeable gain using value at April 1965

Using this method, an asset is assumed to have been acquired on 6 April 1965 at its market value at that time. The difference between disposal price and market value at 6 April 1965 (less any other allowable expenditure) will therefore be the gain in value since April 1965 and consequently the chargeable gain for Capital Gains Tax purposes.

		Sale price	£5 000 000
less	Value in April 1965 [see note 1]	£ 5 000	
1966	Cost of laying tarmacadam	£ 2 500	
1967	Cost of marking out spaces	£ 500	
1969, 1971 and			
1973	Cost of repairs to surface	£ 450	
1986	Costs of planning appeal [see note 2]	£ 2 500	
1988	Costs of development	£1 200 000	£1 210 950
	Gain in value from April 1965		£3 789 050
less	Indexation allowance (as previously calculated)		£ 45 916
	Chargeable gain		£3 743 134

Notes

1: See note 1 of calculation using value at March 1982. The same principle has been applied to the 1965 value of the property.

2: See note 2 of the calculation using value at March 1982.

The chargeable gain of £3 701 584 calculated using the provisions of the *Finance Act 1988* is the lowest of the three and would therefore be adopted as the figure upon which Mr Robinson's Capital Gains Tax liability would be assessed.

Question 4.2

Before Timbo Pet Supplies can be advised of the likely chargeable gain should they dispose of their warehouse, it is first necessary to calculate the gain that resulted from compulsory acquisition of their previous property.

Gain on compulsory acquisition of former premises

	Compensation received		£78 000
less	Disturbance payment [see note 1]		£ 8 000
	Compensation for premises, i.e. disposal price		£70 000
less	Purchase price 1955	£17 000	
	Improvements in 1960	£ 6 000	£23 000
	Gain from 1955 [see note 2]		£47 000

Proportion of gain attributable to [see note 3] :

Purchase   £47 000 x $\dfrac{£17\,000}{£23\,000}$ =   £34 739

Improvement   £47 000 x $\dfrac{£\,6\,000}{£23\,000}$ =   £12 261
expenditure in 1960

Time apportionment [see note 4]

Purchase   £34 739 x $\dfrac{13}{10+13}$ =   £19 635

Improvement   £12 261 x $\dfrac{13}{5+13}$ =   £ 8 855
expenditure in 1960

        Chargeable gain   £28 490

Notes

1: Compensation was received for disturbance, but is payable separately. Only the amount received for the premises is taken into consideration.
2: No allowance for inflationary gains, since provisions for this were not introduced until March 1982.
3: Calculation to separate gains caused by different items of expenditure (see Question **4.1**).
4: Calculation of the gain after 6 April 1965 (see Question **4.1**). Whole years are assumed.

When a property was acquired prior to 6 April 1965, the taxpayer may elect, in place of using the time apportionment method, to have the gain calculated by reference to the actual value of the property in April 1965. However, no information regarding the value of the property in April 1965 is given in the question and it is assumed this calculation is not required.

£28 490 will therefore be adopted as the chargeable gain in 1978.

It should now be explained that since the proceeds of compulsory acquisition were applied to purchase their present premises, Timbo Pet Supplies would have been eligible for what is usually called 'roll over' relief. A brief outline of this relief should be provided in the answer.

When a trader disposes of assets which have been used only for purposes of trade throughout the period of ownership and disposal proceeds are used to purchase replacement assets, which will also be used for trade purposes, then the trader can claim the following:

(a) that the disposal price of the old asset, if more than the original purchase price, was such that neither a gain nor a loss was achieved and
(b) that the cost of the new asset, be reduced by the gain that actually occurred on disposal of the old asset.

In other words, Timbo Pet Supplies could claim that the gain of £28 490 in 1978 would not be charged to tax at that time, but carried forward on to the present premises. It will be taxed when these premises are disposed of and not replaced.

The new premises must have been acquired within 1 year before and 3 years after disposal of the old premises and it is assumed that this applies to the property occupied by Timbo Pet Supplies.

If the present warehouse is sold, it will not be replaced and the gain rolled over in 1978 will become chargeable. The calculations are as follows.

Calculation using market value in March 1982

Proposed sale price		£245 000
less Value in March 1982 [see note 1]	£190 000	
Reduced by rolled over gain	£ 28 490	£161 510
		£ 83 490
less Indexation allowance [see note 2]		

$$£190\,000 \times \frac{134.3 - 100}{100} \qquad\qquad £\ 65\,170$$

Chargeable gain	£ 18 320

Notes

1: Although a leasehold interest is involved, it is not treated as a wasting asset, since the lease has more than 50 years unexpired.
2: Allowance for inflationary gains since 31 March 1982 (see Question **4.1**).

Calculation using provisions prior to Finance Act 1988

Proposed sale price		£245 000
less Purchase price [see note 1]	£155 000	
Reduced by rolled over gain	£ 28 490	£126 510
		£118 490
less Indexation allowance [see note 2]		£ 65 170
Chargeable gain		£ 53 320

Notes

1: If this produces a lower chargeable gain it may be used in place of market value in March 1982 (see Question **4.1**).
2: Value at March 1982 is greater than expenditure up to 31 March 1982, therefore allowance for inflation is based on this value.

The calculation using market value at March 1982 produces the lower chargeable gain and would therefore be adopted.

Timbo Pet Supplies should be advised that the likely chargeable gain if they proceed with the sale at £245 000, is £18 320.

Question 4.3

Goldsmith acquired a 50 year lease in April 1964 and, in effect, is now selling 20 of those years to Sheridan. He acquired the 50 years for £30 000 and is disposing of 20 years for £100 000. It should be explained that for Capital Gains Tax purposes this is a part disposal of Goldsmith's interest. Dealing firstly with the calculation of the chargeable gain using the provisions prior to the *Finance Act 1988*, it is necessary to determine how much of the £30 000 was paid for the 20 years that Goldsmith is sub-letting. Expenditure on improvements is similarly treated.

Leases with 50 years or less unexpired are treated for taxation purposes as wasting assets — for a detailed consideration of this, the reader is referred to chapter 2 of *Applied Valuation* by Diane Butler (Macmillan, 1987).

On disposal, purchase price and improvement expenditure must be written down in accordance with *Schedule 3, Capital Gains Tax Act 1979*. Writing down may be achieved by using the percentages shown in the table to Schedule 3. These percentages are based on the single rate Years' Purchase tables at 6 per cent, therefore the Years' Purchase tables may be used as an alternative. This is the method adopted here.

In outline, the calculation is as follows:

Premium received for sublease	£100 000

less Proportion of premium paid for head lease that is applicable to sublease, calculated by:

$$\text{Premium paid} \times \frac{\begin{array}{c}\text{YP 6 per cent} \\ \text{for unexpired} \\ \text{term of head lease} \\ \text{when sublease} \\ \text{granted}\end{array} - \begin{array}{c}\text{YP 6 per cent} \\ \text{for unexpired} \\ \text{term of head lease} \\ \text{when sublease} \\ \text{ends}\end{array}}{\begin{array}{c}\text{YP 6 per cent for length of head lease when} \\ \text{originally acquired}\end{array}}$$

However, before this calculation can be carried out it needs amending, to take account of the fact that the premium received by Goldsmith and the premium paid by him to Byron are not on the same basis. Goldsmith's premium to Byron was based on him paying a rent of £30 000 per annum, whereas Sheridan's premium to Goldsmith was calculated on the payment of £65 000 per annum in rent. Because of this, the calculation shown above is reduced by the fraction

$$\frac{\text{Premium received}}{\begin{array}{c}\text{Premium that would be received if head}\\\text{lease and sublease rents were the same}\end{array}}$$

Thus, before calculation of the chargeable gain may proceed, it is necessary to calculate the premium that would have been received by Goldsmith if Sheridan paid a rent of only £30 000 per annum.

This may be done in various ways, for example:

(i) Annual equivalent of premium = $\dfrac{£100\,000}{\begin{array}{c}\text{YP 20 years at 12 per cent and 3 per cent}\\\text{(tax 40 per cent) [see note 1]}\end{array}}$

$$= \frac{£100\,000}{5.494} = £18\,202 \text{ pa}$$

Full rental value = £65 000 + £18 202 = say		£83 200 pa.
	Full rental value	£83 200 pa
less	Sublease rent if same as head lease rent	£30 000 pa
	Profit rent	£53 200 pa
YP 20 years 12 per cent and 3 per cent (tax 40 per cent)		5.494
Premium if sublease rent same as head lease rent		£292 281
	say £292 000	

(ii)	Rent paid by Sheridan	£65 000 pa
less	Rent paid by Goldsmith	£30 000 pa
	Difference in rents	£35 000 pa
YP 20 years 12 per cent and 3 per cent (tax 40 per cent)		5.494
	Premium for difference in rents	£192 290
plus	Premium paid by Sheridan	£100 000
Premium if sublease rent same as head lease rent		£292 290
	say £292 000	

Note

1: In the absence of information to the contrary, it is assumed the rent is fixed for 20 years. High remunerative yield is adopted to reflect this.

A further problem that needs to be resolved, before proceeding, is that the unexpired term of the head lease when the sublease is granted (and consequently when it ends) is not an exact number of years. This means that the Years' Purchase (or appropriate percentage if used) needs to be calculated for the exact number of years and months.

Schedule 3 of the Capital Gains Tax Act 1979 provides that if the duration of a lease is not an exact number of years, the appropriate percentage is derived by taking the percentage for the whole number of years, plus $\frac{1}{12}$ of the difference between this and the percentage for the next higher number of years for each odd month.

The same principle is applied if Years' Purchase is used as an alternative.

Assuming, in the absence of information, that all events occur at the beginning of the month in question, then on grant of the sublease, the head lease would have 25 years 1 month unexpired. At the end of the sublease, there will be 5 years 1 month of the head lease unexpired.

Calculation of YP 6 per cent for 25 years 1 month

YP 6 per cent for 25 years   = 12.7834
YP 6 per cent for 26 years   = 13.0032
Difference   = 0.2198

YP 6 per cent for 25 years 1 month = $12.7834 + \left(\dfrac{0.2198}{12} \times 1 \right)$

$$= 12.8017$$

Calculation of YP 6 per cent for 5 years 1 month

YP 6 per cent for 5 years   = 4.2124
YP 6 per cent for 6 years   = 4.9173
Difference   = 0.7049

YP 6 per cent for 5 years 1 month = $4.2124 + \left(\dfrac{0.7049}{12} \times 1 \right)$

$$= 4.2711$$

All the information is now available to proceed with calculation of the chargeable gain.

Calculation using provisions prior to Finance Act 1988

Premium received   £100 000

less

$$£30\,000 \times \frac{\text{YP 6 per cent for 25 years 1 month} - \text{YP 6 per cent for 5 years 1 month}}{\text{YP 6 per cent for 50 years}}$$

$$= £30\,000 \times \frac{12.8017 - 4.2711}{15.7619}$$

$$= £16\,236 \text{ [see note 1]} \times \frac{£100\,000}{£292\,200} = £5\,556 \text{ [see note 2]}$$

$$£12\,000 \times \frac{12.8017 - 4.2711}{14.3106} \quad \text{[see note 3]}$$

$$= £7\,153 \text{ reduced by } \frac{£100\,000}{£292\,200} = £2\,448 \text{ [see note 4]} \qquad \frac{8\,004}{£91\,996}$$

less   Indexation allowance [see note 5]

It is first of all necessary to calculate the proportion of the value of Goldsmith's interest in March 1982 that is applicable to the 20 year sublease.

$$£230\,000 \times \frac{\text{YP 6 per cent for 25 years 1 month} - \text{YP 6 per cent for 5 years 1 month}}{\text{YP 6 per cent for 32 years}}$$
[see note 6]

$$= £230\,000 \times \frac{12.8017 - 4.2711}{14.0840}$$

$$= £139\,310 \text{ reduced by } \frac{£100\,000}{£292\,200} = £47\,676 \text{ [see note 7]}$$

Indexation allowance

$$= £47\,676 \times \frac{132.6 - 100}{100} = \text{[see note 8]} \qquad\qquad £15\,542$$

Indexed gain from 1964   =   £76 454

Time apportionment [see note 9]

$$£76\,454 \times \frac{23\frac{11}{12}}{1 + 23\frac{11}{12}}$$

= £73 386 = indexed gain from April 1965

This is not the chargeable gain, since allowance must be made for Schedule A Income Tax. It is necessary to explain that for taxation purposes, a premium is treated partly as capital and partly as income, therefore the amount chargeable under Schedule A as income is deducted from the total gain to give the chargeable gain.

Calculation of charge to Schedule A Income Tax

Premium received	£100 000
less (20 − 1) x 2 per cent = 38 per cent [see note 10]	£ 38 000
Amount chargeable to Schedule A	£ 62 000

However, since the head lease was granted after 5 April 1963, allowance against this amount is given for the Schedule A Income Tax already charged to Byron on the premium he received for granting the head lease.

Amount upon which Byron was charged Schedule A Income Tax

Premium received	£30 000
less (50 − 1) x 2 per cent = 98 per cent	£29 400
Amount chargeable to Schedule A	£ 600

Goldsmith's charge to Schedule A Income Tax

Amount chargeable, as calculated above	£62 000
less Allowance for amount on which Byron was charged	
$£600 \times \dfrac{20}{50}$ [see note 11]	£ 240
Amount chargeable	£61 760

Thus, chargeable gain for Capital Gains Tax purposes

= £73 386 − £61 760

= £11 626

Notes

1: Proportion of £30 000 premium assumed to have been paid for the 20 years now being disposed of.

2: Premium paid for 20 years of head lease now on same basis as premium received.

3: YP 6 per cent for 33 years 7 months which was the unexpired term of the head lease when the improvements were carried out in September 1980.

4: Improvement expenditure applicable to 20 years being disposed of.

5: Allowance for inflationary gains since 31 March 1982 based on the greater of market value at that date or total expenditure up to that date (see Question **4.1**).

6: Unexpired term of head lease in March 1982.

7: Proportion of March 1982 value assumed to be applicable to 20 year sublease.

8: Market value at March 1982 greater than expenditure up to 31 March 1982, therefore the indexation allowance is calculated by reference to market value.

9: To calculate the part of the total gain that has occurred after 6 April 1965 (see Question **4.1**).

10: Statutory calculation. Two per cent of the premium, for every year of the lease except the first, is assumed to be capital.

11: $\dfrac{20}{50} = \dfrac{\text{duration of sublease when granted}}{\text{duration of head lease when granted}}$.

When adopting the provisions prior to the *Finance Act 1988*, the taxpayer might also consider the chargeable gain using value at April 1965 rather than time apportionment (see chapter 2 of *Applied Valuation* by Diane Butler (Macmillan, 1987)) but the necessary information is not provided in the question. Given that the head lease was granted in April 1964, a reasonable estimate of value in April 1965 might be achieved. However, it is suggested that there probably would not be any appreciable difference in the chargeable gain and it has been assumed that the calculation is not required in this case.

The chargeable gain produced by the time apportionment method should now be compared with the chargeable gain produced when using market value in March 1982.

Calculation of chargeable gain using market value in March 1982

	Premium received		£100 000
less	Market value in March 1982 applicable to		
	sublease [see note 1]	£47 676	
	Indexation allowance [see note 2]	£15 542	
	Amount chargeable to Schedule A		
	[see note 3]	£61 760	£124 978
		Loss	£ 24 978

Notes

1: As previously calculated. Acquistion is assumed to have been at 31 March 1982 for market value at that date, thus considering only the gain in value since then. Improvements in September 1980 may be ignored since they will be reflected in the value at March 1982.

2: As previously calculated, by reference to March 1982 value. There is no comparison with expenditure up to that time because of assumed acquisition in March 1982.

3: As previously calculated.

Calculation under the provisions of the *Finance Act 1988* produces a loss and Goldsmith will pay no Capital Gains Tax on receipt of the £100 000. The loss of £24 978 may be carried forward to set against any future gains made by Goldsmith.

Question 4.4

(a) *Basis of assessment*

It should be explained that the tax applicable in this case is Inheritance Tax and a brief outline of the tax should be provided.

From 1 January 1985, Estate Duty was replaced by Capital Transfer Tax, by the *Capital Transfer Tax Act 1984*. Capital Transfer Tax was itself replaced by Inheritance Tax under the provisions of the *Finance Act 1986*, with effect from 18 March 1986. Under *Sections 100–107 and Schedules 19 and 20 of the Finance Act 1986* the *Capital Transfer Tax Act 1984* was renamed the *Inheritance Tax Act 1984*.

Under Capital Transfer Tax provisions, gifts made during a person's lifetime were taxed at increasing rates, higher rates being applicable as more chargeable transfers were made. Apart from certain exceptions, Inheritance Tax removed the immediate charge on lifetime transfers, but introduced the concept of Potentially Exempt Transfers. Where such a transfer of value is made during a person's lifetime, it is not charged to tax, but liability to pay Inheritance Tax remains until the donor has survived for a further 7 years. If the donor is still alive 7 years after the date of transfer, the potentially exempt transfer becomes fully exempt. When a transfer of value is, or becomes, chargeable, the basis of charge is the reduction in value of the transferor's estate, rather than the actual value of the item or amount transferred. If the donor pays the tax upon the amount transferred, the reduction in value of his estate will be the value transferred plus the tax payable.

On death, the reduction in value of the deceased's estate will be the total value of the estate. The deceased is assumed to have made a transfer of value, equal to the value of the entire estate, immediately before his death. This value is the amount upon which Inheritance Tax is charged.

There is only one scale of tax, which, at the time of writing, is NIL on the first £110 000 of value and 40 per cent on the remainder. Any lifetime transfers that are chargeable are taxed at half these rates. Potentially exempt transfers that become chargeable because of the transferor's death within 7 years, are taxed on the scale applicable at the date of death. It appears from the question that the situation involves only transfer of the estate on the death of the owner.

(b) *Principles of Valuation*

In valuing property for Inheritance Tax purposes, there are various assumptions that have to be made. Some of these assumptions are statutory and some follow legal precedent.

An outline of these assumptions should be given, together with explanations where necessary and a comment on the possible effect in the case in question. The hypothesis, applied to the case in the example, is that the deceased's estate is offered for sale immediately prior to his death and the agreement reached as to price between hypothetical willing seller and buyer, is the value applicable for Inheritance Tax purposes.

Statutory assumptions and legal decisions clarify what is to be assumed and what disregarded in this mythical market. These may be summarised as follows:

(i)   No reduction in value is to be made for a 'flooded market'.

If a person's entire estate were actually placed on the market at the same time, the market might well be 'flooded' by the sudden increase in supply of a particular type or types of property. Increased supply with static demand could result in depressed values. For Inheritance Tax purposes, each individual property must be valued as if it comes on to the market in isolation. The deceased's estate would probably not cause flooding of the market in its entirety, but nevertheless the 3 farms will be valued as if placed individually on the market, as will the 10 houses and 6 shops.

(ii) Any depreciation in value shown to be directly attributable to the transfer may be reflected in the valuation. An example might be the goodwill of a business that was substantially due to the transferor's personality. This is not likely to have any effect on the value of the estate under consideration.

(iii) It must be assumed that the estate is offered on the market in such a way that the largest number of bidders will be attracted and, consequently, the best price achieved. This is referred to as 'prudent lotting' and follows the case of *Ellesmere v. C.I.R. 1918*. In the example, if it could be shown, for instance, that the owner/occupied farm would achieve a better price in the market if the 116 hectares was divided into smaller areas, then this situation would be assumed.

(iv) Although a hypothetical sale of the estate is assumed, hypothetical costs of such a sale are not allowed to be set against the value.

(v)   The property must be valued in its physical condition at the date of transfer.

(vi) The probability of a show of special interest by a particular purchaser must be fully exploited in the valuation where, from the circumstances of the case, it is clear that such a special purchaser would be in the market. For example, an adjoining owner might wish to purchase the 116 hectare farm and be prepared to pay more for it than any other purchaser, and this would be reflected in the valuation. Authority for this stems from *C.I.R. v. Clay* (sometimes referred to as '*the Plymouth Nursing Home case*').

(vii)   Death is often taken by transferees as an opportunity to dispose of property. In *Middleton and Bainbridge v. C.I.R.* it was stated that the price realised on such a disposal must be strong evidence of value on death and only where circumstances clearly differ at the times of death and sale would the sale price not be used for taxation purposes. In the example, the tenants of the houses and shops may express an interest in purchasing the freehold of the properties they occupy. Or the properties could be sold as an investment and, likewise, the tenanted farms.

(c) *Outline valuation of the estate*

Since little information is given in the question, various assumptions will have to be made, bearing in mind that only an outline valuation is required.

(i) *Valuation of the farmland*

It should be noted that special provisions apply to agricultural properties and certain reliefs are available in this case. These should be briefly described before proceeding with the valuation.

Agricultural value is the value of the property subject to a perpetual covenant prohibiting its use for anything other than agricultural purposes.

The farm of 116 hectares owned, occupied and farmed by the deceased, is eligible for 50 per cent relief in value, because the deceased had vacant possession of the property immediately prior to his death. The tenanted farms will be eligible for 30 per cent relief in value, since the deceased had owned the farms for 14 years prior to his death. The minimum period of ownership necessary to qualify for relief in the case of tenanted farms, is 7 years ending with the date of transfer.

Although relief is available on the agricultural value, any value over and above this must be reflected and, in the example, the value with development potential will be taken into account.

Agricultural value:

Owner occupied farm.

116 hectares at £5 200 [see note 1]		£603 200
less Agricultural relief at 50 per cent		£301 600
		£301 600

Tenanted farms

88 hectares at £95 [see note 2]	£8 360 pa		
YP in perpetuity at 4 per cent [see note 3]	25	£209 000	
93 hectares at £95 [see note 2]	£8 835 pa		
YP in perpetuity at 4 per cent	25	£220 875	
Capital value		£429 875	
less Agricultural relief at 30 per cent		£128 963	£300 912
Reduced agricultural value of the 3 farms [see note 4]			£602 512

Value with development potential

116 hectares at £11 000 [see note 5]		£1 276 000
2 tenanted farms [see note 6]		£ 429 875
Value with development potential		£1 705 875
less Agricultural value:		
Owner/occupied farm	£603 200	
2 tenanted farms	£429 875	£1 033 075
Difference in values [see note 7]		£ 672 800
plus Reduced agricultural value		£ 602 512
Value for Inheritance Tax purposes		£1 275 312

say £1 275 300

Notes

1: Value per hectare assumed.

2: Tenanted farms assumed let at current rental value in absence of further information. Rental value per hectare assumed.

3: Yield of 4 per cent assumed reasonable.

4: Agricultural value of the 3 farms after deduction of available reliefs.

5: Assumed value with development potential.

6: Agricultural value. Development potential is applicable only to the farm occupied by the deceased.

7: Value over and above agricultural value upon which no relief is available.

(ii) *Valuation of the 10 tenanted houses*

Total rental income 10 x £30 x 52 [see note 1]		£15 600 pa
less External repairs say 7.5 per cent [see note 2]	£1 170	
Management say 5 per cent	£ 780	£ 1 950 pa
Total net income		£13 650 pa
YP in perpetuity at 10 per cent [see note 3]		10
Capital value		£136 500

Notes

1: Ten houses let at £30 per week for 52 weeks. In the absence of information this is assumed to be rent recently determined for the houses.
2: Estimated amount for external repairs 7.5 per cent of rent received.
3: Yield of 10 per cent adopted as reasonable for the houses in question.

It might be assumed that some or all of the tenants would be interested in purchasing the freehold of their homes. The show of such interest should be actual and not purely speculation by the valuer. If the tenants are interested in purchasing, they are in the position of special purchasers and it is usually anticipated that they would be prepared to pay a higher price than investment value — somewhere between this and vacant possession value. Assuming the vacant possession value of each house to be £35 000 and, from the previous calculation, the investment value is £13 650, the sitting tenant value may be in the region of £20 000 to £25 000. No reference is made in the question to the tenants showing interest in purchasing the houses, and, although the possibility should be referred to in the answer, as above, it will be assumed that it does not apply in this instance.

(iii) *Valuation of the 6 tenanted shops*

Total rent received 6 x £26 500 =	£159 000	
YP 4 years [see note 1] at 5 per cent	3.546	£563 814
Reversion to 6 x £28 000	£168 000	
YP in perpetuity deferred 4 years at 6 per cent [see note 2]	13.202	£2 217 936
Capital value		£2 781 750

Notes

1: Shops were let on basis of 5 year reviews 6 years ago, therefore next review is in 4 years' time. Review is assumed to be to current rental value.
2: All risks yield for this type of property assumed to be 6 per cent. Traditional yield pattern adopted, with term income valued at 1 per cent less than reversion because of assumed extra security.

Finally, a summary of the various valuations should be given as follows:

Agricultural properties	£1 275 300
10 tenanted houses	£ 136 500
6 tenanted shops	£2 781 750
Total value for Inheritance Tax Purposes	£4 193 550

Question 4.5

A residual valuation approach will be adopted to illustrate this situation, which considers the possible effect of Value Added Tax on a developer's profit.

Gross development value [see note 1]			£12 500 000
less			
(i)	Building costs [see note 2]	£3 000 000	
(ii)	Architect's and quantity surveyor's fees say 10 per cent of (i)	£ 300 000	
(iii)	Finance on (i) and (ii) at 16 per cent for 12 months [see note 3]	£ 528 000	
(iv)	VAT on (i) and (ii) at 15 per cent [see note 4]	£ 495 000	
(v)	Estate agent's and legal fees say 3 per cent of gross development value	£ 375 000	
(vi)	VAT on (v) at 15 per cent	£ 56 250	
(vii)	Promotion say	£ 3 000	
(viii)	VAT on (vii) at 15 per cent	£ 450	
(ix)	Contingencies, including finance say	£ 30 000	
(x)	Purchase of site [see note 5]	£4 500 000	
(xi)	Costs of (x) at 4 per cent	£ 180 000	
(xii)	VAT on (x) and (xi) at 15 per cent	£ 702 000	
(xiii)	Finance on (x) and (xi) at 16 per cent for 2 years [see note 6]	£1 617 408	£11 787 108
			£ 712 892
	× PV of £1 in 2 years at 16 per cent		0.743
	Developer's profit		£ 529 679

Notes

1: Sale of a new freehold building after 1 April 1989 is subject to Value Added Tax at 15 per cent. This applies whether or not the developer elects to subject rents on letting to VAT. If there was a contract for sale in writing prior to 21 June 1988, the sale would be zero rated, but it is assumed that this does not apply. In this case, therefore, on sale of the development, VAT would be 15 per cent of £12 500 000 = £1 875 000.

2: Assumed to include site preparation.

3: Finance to cover building costs, architect's and quantity surveyor's fees is assumed to be borrowed for half the development period. Interest on the loan remains exempt from VAT.

4: Costs of construction services became subject to VAT from 1 April 1989. If a written contract for these services existed on 21 June 1988, the costs would be zero rated, but it is assumed this does not apply.

5: Purchase of freehold building land became subject to VAT from 1 April 1989, unless a written contract for sale existed prior to 21 June 1988, in which case the sale would be exempt. It is assumed that such a contract did not exist.

6: Purchase of the land will occur before development, therefore it will need to be financed for the whole of the development period. Interest on the loan remains exempt from VAT.

Having produced the above calculation, consideration should be given to the possible effects of the developer's decision.

Value Added Tax of £1 875 000 will be chargeable when the completed development is sold. If the developer has not elected to charge VAT on the rents when the development is let, he will not be able to reclaim VAT paid by him on the various elements of construction costs. In this case, his profit will be £529 679 or 4.24 per cent of gross development value.

If, however, the developer does elect to subject rents to VAT — such election being available from 1 August 1989 — VAT that he has paid on the costs of construction may be reclaimed. Thus a total of £1 253 700 would be reclaimable, increasing the developer's profit.

	Gross development value	£12 500 000
less	Costs excluding VAT	£10 533 408
		£ 1 966 592
	x PV of £1 in 2 years at 16 per cent	0.743
	Developer's profit	£ 1 461 178

This represents 11.69 per cent of gross development value. Therefore if the developer elects to charge VAT on rents when the development is let, the VAT he pays during the construction period will not affect his profit, but may affect his cash flow.

Although it would appear beneficial to make an election to charge VAT on rents, the developer perhaps ought first to consider the prospective tenants before making a final decision. If a tenant is VAT registered and therefore able to recover VAT paid, then the tenant may well be indifferent to the addition of VAT to the rent. This is likely to be so of tenants of shops in the development. On the other hand, if the tenant is in the financial services sector, such as a bank, unable to recover

VAT paid, then, depending upon relative bargaining positions, the developer may have to concede that the rent paid is inclusive of VAT. Consequently the developer's profit will be reduced. The developer should perhaps consider carefully the choice of tenants for offices in the development.

Although advice is required for the developer, it might be suggested that any prospective tenants would also be wise to seek advice and clarify the exact position before committing themselves.

For instance, the developer may decide not to subject rents to VAT and he then lets the development. The developer may then sell the development to an investor. If this investor has elected to charge VAT on rents, the tenants will find their circumstances changed. They entered into a lease paying no VAT on the rent, whereas with the change of landlord, tenants will find themselves with an added burden not previously anticipated.

Question 4.6

Initially it should be explained that from 1 August 1989, landlords may elect to charge Value Added Tax on the rents from buildings, whether existing or new. The effects upon both landlord and tenant will depend upon not only the type of tenant, but also the bargaining power of the two parties, particularly at the end of the existing lease when a new lease may be negotiated.

During the period up to the end of the respective leases, the effect upon the tenants will depend upon whether or not they are able to recover the whole or part of VAT paid.

Businesses within the financial services sector, such as Insurance Companies, are usually wholly or partially exempt from VAT and therefore unable to recover VAT paid. Charities are in a similar position, since they are exempt.

In the premises in question, should the landlord elect to charge VAT on the rents, during the current leases both the Charity and the Insurance Company will be subject to the increased cost of VAT, unless they decide to sell their interests.

From the landlord's point of view, it may seem desirable to elect for VAT on the rents, because if the services he supplies are charged at an amount including VAT or have VAT added, then VAT on goods and services supplied to him will be recoverable. Thus, VAT on the latter may only affect the landlord's cash flow in the short term. If the tenant is also able to recover VAT on rent then they too will suffer no long-term effect. The tenant's costs will not be increased by the imposition of VAT and, effectively, the amount of rent paid by the tenant will remain unchanged. In this case, neither landlord nor tenant will be affected long term if rents are charged to VAT.

Should the tenant be unable to recover VAT, the addition of the tax to their rent will consequently increase costs. The tenant in this position may have to decide whether or not this increased burden is so significant that alternative accommodation must be sought where VAT will not be charged.

This decision is bound to depend upon the amount of VAT involved, the ability of the tenant to absorb the extra cost and the availability of alternative accommodation. If there is a shortage of accommodation, then the tenant may be forced to accept the imposition of VAT. However, if alternative space is readily available, the tenant may be in a much stronger position and be able to force the landlord to accept that the rent payable together with VAT should not exceed the rent payable if VAT were not imposed. In other words, the landlord would have to absorb costs of VAT on rent.

Simple calculations would illustrate the effect on the landlord in question.

Shop occupied by Charity

Rent received for next 4 years, excluding VAT	£25 000 pa

Rent that would be received if landlord absorbed the VAT:

$$£25\,000 \times \frac{100}{115} =$$

£21 739 pa

plus   VAT at 15 per cent   £ 3 261 pa

Total rent payable by Charity   £25 000 pa

Offices occupied by Insurance Company

Rent received for next 4 years, excluding VAT	£23 000 pa

Rent that would be received if landlord absorbed the VAT:

$$£23\,000 \times \frac{100}{115} =$$

£20 000 pa

plus   VAT at 15 per cent   £ 3 000 pa

Total rent payable by Insurance Company   £23 000 pa

This would reduce the landlord's income over the next 4 years from £68 000 to £61 739 per annum, a reduction of 9.21 per cent per annum.

The second floor offices are not considered above, on the assumption that the Solicitors would be in a position to recover VAT and therefore indifferent regarding the landlord's election to VAT.

Assuming that new leases are negotiated in 4 years' time and, once more, the landlord has to absorb VAT, in respect of the ground floor shop and first floor offices, the effect upon his income will be as follows:

Shop occupied by Charity

Current rental value excluding VAT £32 000 pa

Rent that would be received if landlord absorbed the VAT:

$$£32\,000 \times \frac{100}{115} =$$ £27 826 pa

plus VAT at 15 per cent £ 4 174 pa

Total rent payable by Charity £32 000 pa

Offices occupied by Insurance Company

Current rental value excluding VAT £30 000 pa

Rent that would be received if landlord absorbed the VAT:

$$£30\,000 \times \frac{100}{115} =$$ £26 087 pa

plus VAT at 15 per cent £ 3 913 pa

Total rent payable by Insurance Company £30 000 pa

The landlord's income in 4 years' time will be reduced from £89 500 to £81 413 per annum, a reduction of 9.04 per cent per annum.

The effect upon the tenants if they have to accept imposition of VAT should also be considered. Again, simple calculations will show the increased costs involved.

Charity

Rent payable over next 4 years £25 000 pa
plus VAT at 15 per cent £ 3 750 pa

Total payment £28 750 pa

Insurance Company

Rent payable over next 4 years £23 000 pa
plus VAT at 15 per cent £ 3 450 pa

Total payment £26 450 pa

Solicitors

Rent payable over next 4 years £20 000 pa
plus VAT at 15 per cent £ 3 000 pa

Total payment £23 000 pa

But the total amount payable by the Solicitors will remain at £20 000 pa since VAT will be recoverable. There would be no effect on the net amount received by the landlord, since the amount received less VAT paid to Customs and Excise, will equal the original amount of rent received.

If new leases are negotiated at the end of 4 years and the tenants have to accept the added cost of VAT, total payments by the Charity, Insurance Company and Solicitors will be £36 800 pa, £34 500 pa and £31 625 pa respectively. However, the rent paid by the Solicitors will effectively be £27 500 pa since they will be able to recover the VAT of £4 125 pa.

In order to help tenants, particularly those already in occupation, who suddenly find their costs increased because the landlord elects to charge VAT on rents, certain transitional arrangements are applicable and these should now be considered.

Where VAT is charged, payment will be phased, and, in the case of charities, who may be more affected than most, phasing will be over 5 years. The transitional arrangements are as follows:

On rents due for the period 1 August 1989 to 31 July 1990, VAT will be charged on half the rent.
From 1 August 1990, VAT on the full amount of rent due will be charged.
If the tenant is a charity, VAT on the full amount of rent due will not be charged until 1 August 1993.

In this example, this would have the following effect, assuming the position is being considered at 1 August 1989:

Shop occupied by Charity

From 1 August 1989 to 31 July 1990, rent payable	£25 000 pa
plus VAT at 15 per cent on $\dfrac{£25\,000}{5}$ =	£ 750
Total payment	£25 750
From 1 August 1990 to 31 July 1991, rent payable	£25 000 pa
plus VAT at 15 per cent on $\dfrac{£25\,000}{5} \times 2$ =	£ 1 500
Total payment	£26 500
From 1 August 1991 to 31 July 1992, rent payable	£25 000 pa
plus VAT at 15 per cent on $\dfrac{£25\,000}{5} \times 3$ =	£ 2 250
Total payment	£27 250

From 1 August 1992 to 31 July 1993, rent payable	£25 000 pa
plus VAT at 15 per cent on $\dfrac{£25\,000}{5} \times 4 =$	£ 3 000
Total payment	£28 000
From 1 August 1993, rent payable	£32 000 pa
plus VAT at 15 per cent on £32 000, i.e. full amount of rent due	£ 4 800
Total payment	£36 800

The latter calculation takes no account of rental growth during the next 4 years — the current rental value of £32 000 per annum will no doubt increase during this time with a consequent increase in VAT.

Offices occupied by Insurance Company

From 1 August 1989 to 31 July 1990, rent payable	£23 000 pa
plus VAT at 15 per cent on $\dfrac{£23\,000}{2} =$	£ 1 725
Total payment	£24 725
From 1 August 1990 to 31 July 1991, rent payable	£23 000 pa
plus VAT at 15 per cent on £23 000, i.e. full amount of rent due	£ 3 450
Total payment	£26 450

The tenant of the second floor offices will be similarly treated but, as previously noted, no long-term effect will result since VAT is recoverable.

The landlord's decision regarding election to charge VAT on rents is apparently not simple. The situation in question illustrates some of the points he should consider and various possible implications of his eventual action. However, the 1989 VAT provisions are as yet at a very early stage. It is difficult to be precise and the full impact of the provisions is virtually impossible to forecast.

5 Rating

The rating system has been in existence for several hundreds of years, 1601 being the date from which the present system, as it is now recognised, originated. The *Poor Relief Act of 1601* provided for taxing the occupation of property within a parish in order to support the poor of that parish and, today, rates remain as a tax on occupation.

At present, the rating system is going through a phase of transition. It has been the subject of debate for several years and this debate has mainly been directed towards seeking an alternative method of raising the required revenue. The Conservatives pledged to abolish the rating system, but in 1983, they decided that the advantages of rating as a local tax outweighed its disadvantages and the system remains.

However, changes are to be introduced. At the time of writing, the Valuation Office is undertaking a revaluation in order to produce a new and long overdue Valuation List, in 1990. For the first time, domestic properties will not appear in the Valuation List, since domestic rating is to be replaced by a Community Charge. Various other changes are proposed to the existing system and its implementation, many of which are still at the consultative stage. A uniform business rate is proposed, to replace the separate rate poundages of individual local authorities and phasing arrangements to allow occupiers to come to terms financially with the effects of the uniform business rate and the increases in liability that would result from the extended time span since the last revaluation in 1973. Inevitably, as with any change, there will be those who benefit and those who are placed at a disadvantage.

Whatever the changes eventually introduced and the results that ensue from the 1990 revaluation, the principles of rating valuation remain, and this section explores both these principles and their application.

Statute is bound to have played a large part in the development of a system that has existed for several centuries. From statute springs the inevitable case law, of which there are volumes concerning rating and valuation for rating purposes. Wherever possible, in answering examination questions, the relevant statutory or case law authority should be quoted.

The questions in this section are arranged so that initially the principles of rateable occupation are considered, followed by the hypothesis surrounding the rating valuation. Methods of valuation are then explored and, finally, some examples of their application. These examples are all assumed to refer to valuations for the 1973 Valuation List.

RATING – QUESTIONS

5.1. Your client is a bank which has recently opened a branch on the concourse of an airport. The premises are held on an annual tenancy and the bank has complete control of the premises but can only gain access to them when the airport is open, except by special arrangement.

The bank has been separately assessed for rating purposes and your client does not feel that this is correct. Advise your client.

5.2. (a) Describe the bases of assessment set out in *Section 19 of the General Rate Act 1967* and explain their application to various types of property in the 1973 Valuation List. Consider how this will change in future Valuation Lists.

(b) Discuss the significance of the doctrine of *rebus sic stantibus* in rating law, having regard to relevant cases, under the following heads:

(i) mode of use of the hereditament
(ii) structural alterations to the hereditament
(iii) state of repair of the hereditament.

5.3. (a) Explain what is meant by valuation according to the 'tone of the list' and discuss why the tone of the list provisions were originally enacted.

(b) In applying the tone of the list provisions, would a public house or mineral hereditament be dealt with any differently from a cinema or a petrol filling station?

(c) Consider whether there are some property value changes which do not fall within the tone of the list provisions for the current Valuation List.

Illustrate your answer throughout with examples and case references.

5.4. Outline the three methods of valuation used to prepare rating assessments, using examples as appropriate. Discuss the strengths and weaknesses of each method, indicating the types of property to which each method may be applied.

5.5. Your client is proposing to take a lease of a ground floor shop, No. 7 Little Street, at a rent of £32 000 per annum. He will be responsible for internal repairs and insurance and will be required to pay a service charge to the landlord in respect of the cost of external repairs to the property. Your client will also have to fit out the shop. The shop has a frontage of 6 m and a depth of 18 m.

In 1972, No. 11 Little Street was let for £4 250 per annum, the tenant being responsible for internal repairs and insurance. This property is also a ground floor shop and now measures 6.5 m frontage by 18 m depth. The shop previously had a rear extension, 5.5 m wide and 4 m deep, which was built in 1966, but demolished in 1980.

Advise your client as to the probable rating assessment of the shop he is proposing to occupy. Show the calculations necessary to arrive at your decision, stating clearly any assumptions that you make.

5.6. You have received instructions from the occupier of a shop in the high street of a provincial town, to check the rating assessment of his premises, which is gross value £3 000 rateable value £2 472.

Your client is the freeholder and the shop measures 6 m frontage by 18 m depth.

As a result of your investigations, you obtain the following information in respect of adjoining shop premises:

Shop 1 Measures 6 m frontage by 16 m depth. It was let in 1972 on a 21 year full repairing and insuring lease at a rent of £1 700 per annum. This lease was granted on the surrender of a previous one, on similar terms, then having 3 years unexpired at a rent of £1 000 per annum, no premium having been paid. The new lease in 1972 was subject to reviews at the end of the 7th and 14th years.

Shop 2 Measures 8 m frontage by 15 m depth. This shop was let on a 14 year full repairing and insuring lease from 1972, with no rent review. The rent passing was £2 000 per annum and, on entry, the tenant paid a premium of £5 000 and agreed to carry out repairs costing £2 000.

The remaining shops in the street are all the same size. Each has a frontage of 5 m and a depth of 18 m and is assessed for rating purposes at £2 225 gross value.

Analyse the foregoing and advise your client on the appropriate action he should take.

5.7. Your client, who is the lessee/occupier of a licensed hotel in the main street of a provincial town, has instructed you to check the rating assessment of £7 250 gross value £6 014 rateable value of his premises. The following information has been extracted from his accounts, averaged over the last 3 years:

Receipts: Restaurant	£47 500
Bars	£71 000
Letting of rooms	£32 500
Purchase of consumable stock	£55 250
Expenses:	
Wages, salaries and National Insurance	£36 000
Insurance (buildings)	£ 350
Insurance (contents and 3rd party)	£ 240
Laundry and cleaning	£ 3 200
Advertising, stationery etc.	£ 890
General rates	£15 035
Postage and telephone	£ 545
Repairs to buildings	£ 3 780
Repairs to furniture and fittings	£ 2 800
Interest on Bank loan	£ 1 500
Occupier's drawings	£ 7 000
Accountant's fees	£ 375
Bad debts	£ 175
Rent	£10 000
Lighting and heating	£ 3 240
Annual sinking fund to replace tenant's chattels	£ 1 025

The furniture and equipment were recently valued at £45 000 but their replacement value is £77 000. The lessee holds a cash float of £6 000 and the average value of consumable stock is £12 000. Rates are 292p in the £.

Prepare a valuation for rating purposes and advise your client on the appropriate action he should take.

RATING – SUGGESTED ANSWERS

Question 5.1

The points that have to be established in this instance are whether or not:

(a) the bank premises is a separately rateable hereditament
(b) your client is the rateable occupier.

In considering these matters, relevant case law should be quoted, as appropriate.

(a) *Is the bank premises a separately rateable hereditament?*
There are several tests which must be applied in identifying a separate hereditament. In this case there should be no problem in identification, but the following points should be referred to briefly in answering the question:

(i) The hereditament can only be occupied for a single function

There can be little argument that the single function for which the premises are occupied is that of providing a banking service.

Relevant cases to quote include:

Eastern Electricity Board v. Smith (VO) and City of St Albans 1974
Poor Sisters of Nazareth v. Gilbert (VO) 1983

(ii) The hereditament must be within a single curtilage
This, again, is likely to be a matter of fact, and the premises are unlikely to be fragmented.

Cases to refer to include:

Gilbert (VO) v. Hickinbottom 1956
Birmingham Roman Catholic Archdiocesan Trustees v. Stamp (VO) 1974

(iii) The hereditament must be within one rating area
This is most likely to be the case, but an airport, usually covering a large area, may stretch over the boundaries of more than one rating area and there is then a possibility that the bank may be so situated on the airport concourse that it straddles such a boundary. If this were so, then the rating assessment would have to be apportioned between the rating areas concerned. Again, this would be a matter of fact.

(iv) The hereditament must be in a definable location
There should be no problem in this respect — the location of the bank on the airport concourse will be obvious.

(v) For each single hereditament, there can be only one rateable occupier
This is the contentious point and leads on to

(b) *Is your client the rateable occupier?*

Consideration must now be given to the essential elements of rateable occupation. Case law, rather than statute, has established that, in order to be rateable, an occupation must be

(i) actual
(ii) exclusive
(iii) beneficial and
(iv) permanent.

Every one of these ingredients must exist in order that an occupation be rateable, and, in the answer, each should be considered in turn, to determine whether or not they are applicable to the bank.

(i) Actual occupation
The premises must actually be in use, no matter how slight that use is. In this instance, there can be little argument that the premises are being used as the branch of a bank and this does not appear to be in dispute.

(ii) Exclusive occupation
If rateable occupation is established, this really is the ingredient that will determine

who the rateable occupier is. For rating purposes, an occupation will be exclusive if it excludes all other persons from using the property in the same way, even though it does not exclude them from using the property in another way.

Relevant case law includes:

Hollywell v. Halkyn Mine Co., 1895

Peak (VO) v. Burley Golf Club, 1960

The latter case involved a golf club using common land in the New Forest, under licence. However, the public were not excluded from the land and, in fact, used the golf club without payment. It was held that the golf club was not in exclusive occupation and therefore not rateable.

Problems usually occur where there is more than one possible occupier of a particular property, which is the case with the bank on the airport concourse. Similar circumstances were considered by the House of Lords in *Westminster Council v. Southern Railway Co. Ltd. 1936*, and the decision should provide a solution to the problem of the bank. It is therefore essential that the case be quoted in answer to this question. The 'Westminster' case concerned various properties let by the Railway Company. These properties, all on London Victoria Station, included a bookstall, shops, kiosks and sites used by coal merchants and builders merchants. There were four things common to the various tenancy agreements:

1. there was a defined site
2. rent was payable to the Railway Company
3. if the premises were to be separately assessed for rating purposes, the occupiers would pay the rates
4. access to the premises was under the control of the Railway Company and the properties could only be used during specified hours, except by special arrangement with the Company.

The circumstances were therefore very similar to those applying to the bank on the airport concourse. In the 'Westminster' case, the tenants of the individual properties were all held to be rateable occupiers and it was stated:

> ". . . the question must be not who is in paramount occupation of the station within whose confines the premises in question are situated, but who is in paramount occupation of the premises in question."

Following this decision, the bank is clearly in exclusive occupation of the branch on the airport concourse. No other person or body is sharing in the use for which the bank holds the premises. Their use is paramount for that purpose, to the exclusion of all others.

(iii) Beneficial occupation

The occupation must be of value to the occupier, but this does not mean that the occupier must make a financial profit, although it would be surprising if the bank did not expect to do so. Basically, for an occupation to be beneficial, it must be such that a hypothetical tenant would be prepared to pay a rent for the right to occupy. It is a matter of fact that the bank in question is paying a rent in order to occupy the premises and they would have difficulty in establishing that their occupation is not beneficial.

A relevant case is *Jones v. Mersey Docks, 1865.*

(iv) Permanent occupation

This means that the occupation should not be too transient in nature and there must be expectation that the occupation will continue for a reasonable period of time.

Relevant case law includes:

L.C.C. v. Wilkins (VO) 1954, where contractor's huts were held to be rateable after being in place for 12 months.

Field Place Caravan Park v. Harding (VO) 1966. Caravans on prepared sites were held to be rateable, also after being in place for more than 12 months.

It appears a reasonable expectation that the bank on the airport concourse will be occupied on a permanent basis. The occupation is likely to be of a more permanent nature than either contractor's huts or caravans and, furthermore, the premises is held on an annual tenancy.

In conclusion, your client should be advised that the occupation by the bank satisfies all the criteria of rateable occupation and it appears that the separate assessment for rating purposes is correct.

Question 5.2

(a) The bases of assessment set out in *Section 19 of the General Rate Act 1967* are Gross Value and Net Annual Value and definitions of these should be provided.

Gross Value is defined in *Section 19(6) of the 1967 Act* as:

> "the rent at which the hereditament might reasonably be expected to let from year to year if the tenant undertook to pay all the usual tenant's rates and taxes and the landlord undertook to bear the cost of repairs and insurance and other expenses, if any, necessary to maintain the hereditament in a state to command that rent."

Net Annual Value is defined in *Section 19(3)* as:

> "an amount equal to the rent at which it is estimated the hereditament might reasonably be expected to let from year to year if the tenant undertook to pay all usual tenant's rates and taxes and to bear the cost of the repairs and insurance and other expenses, if any, necessary to maintain the hereditament in a state to command that rent."

Thus, the difference between the two bases may briefly be summed up as follows:

Gross Value —the landlord is assumed to pay the cost of repairs and insurance necessary to maintain the property in a condition to command the rent.

Net Annual Value —the tenant is assumed to pay the cost of the necessary repairs and insurance.

Rates are charged at so much in the £ (determined each year by the Rating Authority) on the Rateable Value of a property and the rateable value is either derived from gross value or net annual value. In practice, net annual value and rateable value are the same, but net annual value is derived directly whereas rateable value is derived from gross value. If a property is assessed to gross value, there is a statutory deduction from the gross value, allowing for the annual costs of repairs, maintenance and insurance, to derive the appropriate rateable value. Put simply, gross value, reduced by the statutory deduction, produces the rateable value, which is equivalent to net annual value.

The statutory deductions are provided in the *Valuation (Statutory Deductions) Order 1973*. These are shown below, although they would not normally be required in an examination answer.

Gross Value	*Statutory deduction to produce rateable value*
Not exceeding £65	45 per cent of gross value
Exceeding £65 but not exceeding £128	£29 + 30 per cent of amount by which gross value exceeds £65
Exceeding £128 but not exceeding £330	£48 + $\frac{1}{6}$ of amount by which gross value exceeds £128, with a maximum of £80
Exceeding £330 but not exceeding £430	£80 + $\frac{1}{5}$ of amount by which gross value exceeds £330
Exceeding £430	£100 + $\frac{1}{6}$ of amount by which gross value exceeds £430

The next item to consider, is the decision regarding whether a property is assessed to gross value or directly to net annual value.

Although there may occasionally be an element of doubt, reference to statute should determine the correct basis in the majority of cases encountered by a valuer.

Hereditaments to be valued to gross value are defined in *Section 19(2) of the General Rate Act 1967* as:

"consisting of one or more houses or other non-industrial buildings, with or without any garden, yard, court, forecourt, outhouse or other appurtenance belonging thereto but without other land."

All other hereditaments are valued to net annual value. In other words most hereditaments in the 1973 Valuation List will be valued to gross value, except industrial property and those properties where the greater part of the value is in the land.

A few examples would help to clarify the answer.

The following will be valued to gross value:

Shops, hotels, offices, warehouses, houses, schools, public houses.

The following will be valued to net annual value:

Factories and oil refineries, because they are industrial hereditaments.

Golf courses, race courses and quarries, because the value is mainly in the land.

At the next revaluation in 1990, net annual value will be the general basis of assessment. This will be more realistic, since the majority of properties are now actually let on, or near to, this basis. The change stems from *Section 29 of the Local Government, Planning and Land Act 1980*, which provided that dwelling houses, private garages and private storage used in connection with a dwelling house, should be the only property valued to gross value at the next revaluation. Dwelling houses will not be included in the 1990 revaluation, since domestic rating is to be replaced by a Community Charge, therefore net annual value will be the normal valuation basis.

(b) A brief explanation of the doctrine of *rebus sic stantibus* should first of all be given.

Rebus sic stantibus means that, for rating purposes, a property must be valued in its actual physical state, as it stands, at the date of valuation, bearing in mind its benefits, disabilities and its present use.

Consideration may now be given to the interpretation of the doctrine under the heads in question.

(i) *Mode of use of the hereditament*

The property must be valued in its present mode of use at the time of valuation.

Two particular cases are useful to quote in support of this. Firstly, in *N and SW Railway Co. v. Brentford Union 1888*, it was stated that

"the thing must be valued as it is for the purpose for which it is used."

Also, in *R. v. Everist 1847*, it was pronounced unreasonable that property used for one purpose should be considered in terms of an alternative use, since the different uses would produce different levels of profit and different levels of rent.

What must always be borne in mind is that the valuer is seeking to determine the hypothetical rent that a hypothetical tenant would be prepared to pay for the hereditament. This is bound to be affected by the use of the hereditament.

Valuing a property as it is actually used does not mean that a particular type of that use must be considered. A shop will be valued as a shop and not a specific type of shop — no differentiation would be made between say a butcher's and a florist's.

(ii) *Structural alterations to the hereditament*

Rebus sic stantibus assumes that the property is valued in its actual physical state at the date of valuation. It follows from this that the effect of possible and significant structural alterations must be ignored, in the same way as any alternative use of the hereditament is disregarded.

A case which may usefully be quoted in illustration is *Prince v. Baker (VO) and Widnes B.C. 1973*.

This case concerned a shop, together with living accommodation, situated in a parade of shops. The shop was used exclusively as living accommodation and the ratepayer contended that it should be assessed as such for rating purposes. However, it was held that, since there had been no structural alterations, the property should be assessed as a shop. A tenant, coming fresh upon the scene, would be prepared to pay a rent for the premises reflecting its ability to be used as a shop.

(iii) *State of repair of the hereditament*

The first point to make under this heading is that, since the doctrine of *rebus sic stantibus* requires the actual physical state of the hereditament to be considered, it might appear that the state of repair ought to be taken into account. If this were necessary, it would result in frequent reviews of assessments, reflecting each hereditament's current state of repair.

Reference should now be made to the statutory definitions of gross value and net annual value (see part (a) of this question), which fortunately mean that the valuer may generally disregard the state of repair of a hereditament without violating the *rebus sic stantibus* rule.

The statutory assumptions regarding repair should briefly be reiterated. Gross value assumes that the landlord keeps the property in a state of repair necessary to maintain its rental value and net annual value assumes that this is the tenant's responsibility.

Therefore, whatever the actual state of repair, rating hypothesis assumes that each property is in a reasonable state of repair, such that it would command the rent which a hypothetical tenant would pay, bearing in mind both the type of property and the neighbourhood in which it is situated. This would be so unless the state of disrepair was excessive, that is, unless it would be unreasonable to expect the hypothetical landlord (gross value) or the hypothetical tenant (net annual value) to reinstate the premises to a reasonable state of repair. Excessive disrepair would be reflected in the rent that a tenant would be prepared to pay for a property and it should also therefore be reflected in the rating assessment.

An illustrative case to quote in this instance would be *Saunders v. Maltby 1976*.

Question 5.3

(a) In answering this part of the question, it should be noted that the tone of the list of provisions are contained in *Section 20 of the General Rate Act 1967* and concern the valuation of properties for rating purposes.

It should then be explained that *Section 20* provides that after a Valuation List has come into force, when a new property is entered in that list, or an existing property is revalued for some reason, the value placed upon such property must not exceed the value that would have been ascribed to it had it existed in the year prior to the Valuation List coming into force. However, it must be assumed that certain factors affecting the value of the property would have been the same during the year before the Valuation List came into force as they are at the time of valuation. These factors are:

(i) the state of the hereditament and the way in which it is occupied;
(ii) the locality in which the hereditament is situated, including the occupation and use of other premises in the locality, transport services and other factors affecting the amenities of the locality.

The tone of the list provisions thus ensure that all properties of a similar type are valued on a uniform basis throughout the Valuation List. In effect, the tone of the list sets a maximum level of value.

A relevant case to quote in this instance is *K Shoe Shops Ltd. v. Hardy (VO) 1983*, when the House of Lords confirmed that valuation should be at the date of the coming into force of the Valuation List.

Discussion of the reasons for enactment of the tone of the list provisions should now follow.

Essentially, the provisions were intended to avoid the injustice which might otherwise occur in an inflationary economy. During periods of inflation, if new properties entered in the Valuation List were valued on the basis of current rental values, rateable values, and, consequently, rate liabilities, would be higher than similar properties already in the Valuation List. At the time of writing, the Valuation List came into force on 1 April, 1973 and properties will thus be valued on the basis of rental values during the year prior to that date.

Before enactment of the tone of the list provisions, relevant case law would have created unfairness between assessments. For example, *Barratt v. Gravesend A.C. 1941*, established the rule that a hereditament should be valued at the date of proposal, or the date of revaluation if a new Valuation List was being prepared. In inflationary times, the value at the date of proposal could be substantially higher than the value at the date of the Valuation List. In *Ladies Hosiery and Underwear v. West Middlesex A.C. 1932*, it was held incorrect to assess new properties on the same basis as existing properties. Even though it was accepted that this may create unfairness in rate liabilities, the first consideration was held to be uniform correctness of rating assessments.

To overcome the problems thus created in a period of high inflation, the tone of the list provisions were enacted, originally, in the *Local Government Act 1966* and subsequently re-enacted in *Section 20 of the General Rate Act 1967*.

(b) Initially, it should be pointed out that the rental values of all the four types of property mentioned in the question will be influenced by the profit making capabilities of those properties.

Thus, the rating assessments will either be derived via the profits method (see Question 5.4), or some quasi profits method.

Public houses and mineral hereditaments are specifically referred to in *Section 20 of the General Rate Act, 1967,* whereas cinemas and petrol filling stations are not.

Section 20(i) (a) refers to the assumption that, at the date of valuation, the hereditament was in the same state as it would have been had it existed in the year prior to the Valuation List coming into force, together with other "relevant factors", which are defined in *Section 20(2)* as:

"(a) the mode or category of occupation of the hereditament;
  (b) the quantity of minerals or other substances in or extracted from the hereditament; or
  (c) in the case of a public house, the volume of trade or business carried on at the hereditament."

Section 20(3) goes on to state:

"References in this section to the time of valuation are references to the time by reference to which the valuation of a hereditament would have fallen to be ascertained if this section had not been enacted."

In other words, the volume of trade achieved in a public house and the amount of mineral extracted in the case of a mineral hereditament will be taken at the date of valuation. The levels of value applied to the volume of trade or amount of mineral extracted will, however, be those applicable at the date of the Valuation List.

Thus, in the case of a public house, the total barrelage will be calculated at the date of valuation, but the price per barrel will be the value that would have been placed upon it at the date of the Valuation List. Similarly, the volume of mineral extracted from mineral workings will be ascertained at the date of valuation, but the value applied to that volume will be that applicable at the date of the Valuation List. Although not mentioned in *Section 20*, cinemas and petrol filling stations are dealt with in a similar way. Attempts have been made to value cinemas on the basis of a price per seat, for example, *Provincial Cinematograph Theatres Ltd. v. Holyoak (VO) 1969.* In this case, the rate payers contended for a valuation using a price per seat, but the Lands Tribunal adopted the Valuation Officer's basis of gross receipts. There are difficulties involved in using the number of seats as a basis, which may cause anomalies between assessments. Cinemas are often tied to particular film distributors or are part of a chain, therefore comparison with other cinemas in different circumstances may be unsatisfactory. It should be explained that the method adopted is to calculate gross receipts, to which a percentage is applied, depending upon the number of full houses likely to be achieved each week. Gross receipts will be those at the time of valuation, whilst the percentage applied to derive gross value will be on the basis of percentages applicable at the date of the Valuation List, in order to conform with the tone of the list.

Relevant case law includes *Paul Raymond Organisation Ltd. v. Pirie 1979.*

Petrol filling stations are valued using throughput.

In the case of *Petrofina (Great Britain) Ltd. v. Dalby (VO) 1967*, the Valuation Officer argued for the use of the contractors method, but, despite the Lands Tribunal's agreement that it was a valid method, the decision was in favour of basing the valuation on throughput. The total throughput of petrol and derv is valued using a price per 1000 gallons, adding to the resultant figure the value of the buildings and any rateable plant. Throughput adopted will be that at the time of valuation, but the price per 1000 gallons will be that applicable at the date of the Valuation List, the buildings and rateable plant also being valued using levels of value applicable at that date.

In summary, it may be noted that, although the units of comparison employed in the valuation of the four types of property mentioned in the question do differ, the tone of the list provisions are nevertheless applied in a similar way.

(c) Reference should be made in the answer to the case of *Clement (VO) v. Addis Ltd. 1988*.

The word 'state' in *Section 20(1) (a) of the General Rate Act 1967* (referring to the state of the hereditament at the date of the valuation) has usually been taken to mean the physical factors or factors affecting the physical enjoyment of a hereditament, not extending to intangible factors, economic and financial matters, planning decisions or technological changes. This was challenged by the ratepayers in the 'Addis case'. A brief outline of the case will help with the explanation. Addis Ltd. were occupiers of a factory outside, but within one mile of the Lower Swansea Valley Enterprise Zone. Both the valuation officer and the ratepayers agreed that establishment of the Enterprise Zone had resulted in the dpression of rental values outside the Zone, because of the attractiveness of locating within the zone because of fiscal, administrative and other advantages. The parties had also agreed upon the reduced rateable value of the subject factory, if the consequences of the Enterprise Zone were to be taken into account. The matter in dispute was whether or not these consequences ought to be taken into consideration under the provisions of Section 20.

The ratepayers contended that Section 20 encompassed not only physical factors, but also any advantages or disadvantages resulting from primary or subordinate legislation. In support of this contention, reference was made to the decision in *Port of London Authority v. Orsett Union Assessment Committee 1920*, in which Lord Buckmaster said:

"The actual hereditament of which the hypothetical rent is to be determined must be the particular hereditament as it stands, with all its privileges, opportunities and disabilities created or imposed either by its natural position or by the artificial conditions of an Act of Parliament."

It was also pointed out that in *Dawkins v. Ash Brothers and Heaton Ltd. 1969*, a demolition order, made under statutory powers, had been held to be a relevant factor in the valuation. Similarly, therefore, the state of the locality "so far as concerns the other premises situated in the locality" should include the advantages bestowed upon those premises by designation of the Enterprise Zone under statutory powers. These advantages had reduced the value of the appeal property, situated just outside the Zone.

The ratepayers' appeal was allowed. In his speech, Lord Keith of Kinkel said:

"I would therefore hold that the existence of these advantages is an intangible factor which goes to the state of the locality as regards other premises situated there. One effect of the circumstances affecting those premises is to reduce the value of the hereditaments in the same locality which do not enjoy the same advantages because they are not in the Enterprise Zone. I am therefore of the opinion that on a proper construction of *Section 20* the effects of the enterprise zone on the value of the ratepayers' factory are properly to be taken into account."

The 'Addis' decision thus extended the scope of the application of *Section 20*, which would undoubtedly have had wide ranging implications, allowing factors hitherto not considered relevant to be taken into account in the valuation of property for rating purposes, such as the designation of enterprise zones and urban development corporations, the effect of planning decisions and technological changes.

However, it should be noted that, following the decision in the 'Addis case' in the House of Lords, on 11 February 1988, the Environment Secretary, Nicholas Ridley, announced legislation to effectively reverse the decision. With effect from 10 March, 1988, "proposals to amend current rateable values will be determined according to the law as it was understood prior to the decision in the Addis case." The 'Addis' decision would, however, be taken into consideration where proposals were already outstanding.

Considering the number of contested rating assessments that this announcement must have prevented, there was no doubt relief in the Valuation Office. However, the valuer must feel some regret as an individual to see a decision of the highest court in the land overturned.

Question 5.4

What must be remembered is that, no matter which method of valuation is adopted, it is the rental value of the property that is ultimately required.

The valuer has to consider what rent a hypothetical tenant might reasonably be expected to pay for the property on the statutory terms of either gross value or net annual value (see Question 5.2). In the answer it should be noted that the three main methods of valuation used for rating purposes are

1. the rental method
2. the contractors method and
3. the profits method.

There is a fourth method, but it can hardly be described as a valuation method, since a statutory formula is provided. However, this will only be applicable in a minority of cases, such as gas or water undertakings.

For each type of hereditament, a particular valuation method, will generally be used, but this should not be adhered to dogmatically. Reference should be made to the case of *Garton v. Hunter (VO) 1969*, in which the following statement was made which sums up the situation very well:

"we do not look on any of these as being either a 'right' method or a 'wrong' method of valuation; all three are legitimate ways of seeking to arrive at a rental figure that would correspond with an actual market rent on the statutory hypothesis, and if they are properly applied all the tests should in fact point to the same answer; but the greater the margin of error in any particular test, the less is the weight that can be attached to it."

Individual consideration should now be given to the three main methods of valuation.

1. *The Rental Method*

The first essential point to make is that this method is preferable to any other method, where its use is possible, since it provides the best evidence in deriving either the gross value or net annual value of a property. Rental evidence may be either direct — the rent actually being paid for the subject hereditament — or indirect — evidence of the rents achieved from similar properties. Indirect evidence may also be obtained from the rating assessments of similar properties, since they may be taken as an admission by the Valuation Officer of the value of these properties.

When the rental method is adopted, care must be taken that the evidence is either on the same basis as the statutory definitions of gross value or net annual value, or, if not, it is converted on to that basis (see Question **5.2**).

The rental evidence may not accord with the statutory basis for a variety of reasons, even though they may appear to be on the same terms. Examples of these should be given, and might include:

(i) the rent is not at arms length, for example, due to some connection between landlord and tenant,

(ii) goodwill is reflected in the rent,

(iii) the rent was fixed under the provisions of the *Landlord and Tenant Acts* and may, for instance, exclude the value of improvements undertaken by the tenant with the landlord's consent.

(iv) a premium has been paid, or the tenant has undertaken improvements to the property, as a condition of the lease. In this case, the annual value of these capital sums must be added to the rent. A case in point is *Edma (Jewellers) Ltd. v. Moore (VO) 1975*.

In addition to the above, the rental evidence may not be on the same terms as the statutory definitions of gross or net annual value. Examples of the reasons for this should be given, and will include:

(i) repairing obligations of the landlord or tenant may not be in accordance with the statutory definitions,

(ii) the rent may not be on a tenancy from year to year as assumed by statutory definitions.

(iii) the rent may include an amount for services provided by the landlord. If this is so, the amount attributable to the value of the landlord's services must be deducted from the rent and in the absence of information, a decision must be made as to what proportion of the rent is paid for these services. An illustrative case is *Bell Property Trust Ltd. v. Hampstead A.C. 1949*.

It is relevant to note that the majority of rating valuations are carried out using the indirect rental method. In such cases, rental evidence is translated into a value per square metre, which may then be applied in the valuation of other properties. Similarly, when using the comparison of existing rating assessments, the gross value or net annual value may be expressed as a value per square metre and then applied in the valuation of other hereditaments.

A useful case to quote in this instance is *Lotus and Delta Ltd. v. Culverwell (VO) and Leicester City Council 1979*, which provides guidance regarding the use of direct or indirect rental evidence.

This guidance may be summarised as follows:

(i) if the subject property is let, the rent being paid is an acceptable starting point in arriving at the assessment,

(ii) the closer to the statutory definitions of gross value or net annual value that the rent and the terms upon which it is paid are, the more reliable it is as a guide to the assessment,

(iii) rental values of similar properties should be considered and compared with the subject property,

(iv) assessments of similar properties in the Valuation List should be taken into consideration because they may be assumed to be the Valuation Officer's opinion of value of those properties,

(v) although assessments of similar properties should be borne in mind where there is no other rental evidence, it is difficult to disregard the rent actually being paid for the subject property.

It should be stressed that the rental method is the preferred method of valuation, but care is needed in its application. Rental evidence must either be on the required statutory basis or converted to accord with that basis.

Examples should be provided of properties that are most likely to be valued by the rental method and these would include shops, offices, warehouses and dwelling houses (although the latter will not appear in future Valuation Lists). Examples of the application of this method are provided in Questions 5.5 and 5.6.

2. *The Contractor's Method*

This method would only be adopted in circumstances where the valuer is unable to use the rental method. Case law includes *Robinson Brothers (Brewers) Ltd. v. Boughton and Chester le Street 1937*, in which it was stated:

> "Where better evidence is in the circumstances of a particular hereditament impossible, resort may be had to either capital value or cost of construction either of which can, with appropriate corrections, be converted into approximately equivalent terms of annual value."

Thus, the contractor's method may be used when direct or indirect rental evidence is not available. This will involve properties which rarely, if ever, are available to let in the market.

The basic assumption underlying the method should be noted, followed by an explanation of how rental value is derived.

It is assumed that the hypothetical tenant would calculate their rental bid on the basis of the annual cost of providing the property. In other words, the occupier may borrow capital to buy the land and build the property, and the annual interest on borrowed capital will roughly equate to the rental value of the property. The method involves calculating the effective capital value of the property, a percentage of this being taken as rental value.

Effective capital value may be calculated in either of the following ways:

(i) Cost of replacing the existing building

This cost would include any fees and an allowance would be made for age and obsolescence of the existing building. Any unnecessary ornamentation or embellishments are disregarded in the replacement cost, since a hypothetical tenant could not be expected to pay extra rent for these features.

(ii) Cost of providing a simple substitute building

This alternative entails consideration of the cost of providing a modern building that would perform the same function as the subject property.

The cost, derived by either of these two methods, is added to the value of the land of the subject hereditament. Land is valued in its present use, ignoring any potential for alternative development. This follows the rule that a property, for rating purposes, must be valued *rebus sic stantibus*, that is, in its physical state and existing use at the time of valuation (see Question 5.2(b)).

The resultant figure is referred to as effective capital value and this is decapitalised at an appropriate percentage to determine rental value. The percentage adopted will vary according to the type of property involved and whether assessment is to gross value or net annual value.

Guidance is usually obtained from case law and percentages adopted in the 1973 Valuation List include:

Oxford Colleges	$2\frac{1}{2}$ per cent to NAV
Public swimming pool	3 per cent to NAV
Public school	$3\frac{1}{2}$ per cent to GV

The question requires examples to be provided of the types of property which may be valued by this method and the above give some indication. Others that might be mentioned include crematoria, town halls and sewage works.

It is a good idea to construct an example to illustrate the method – a fairly simple example would be sufficient, such as the one shown below.

Example 1

Assessment of public school [see note 1].

Estimated current replacement cost of buildings [see note 2]		£3 500 000
Deduction for age and obsolescence, say 60 per cent		£2 100 000
		£1 400 000
Add:		
6 hectares of sports ground at £4 000 per hectare	£24 000	
2.5 hectares of land covered by buildings, at £12 000 per hectare [see note 3]	£30 000	
Roads, car parks etc. say	£ 7 500	£ 61 500
		£1 461 500
Effective capital value		
at $3\frac{1}{2}$ per cent [see note 4] =	£51 152	
say	£51 150 gross value.	

Notes
1: This is an imaginary public school and all figures are assumed.
2: Deduction depends upon the facts of the case, including the age of the buildings.
3: Value of the land for use as public school. Any alternative use is ignored. The market value of the land may be more but, *rebus sic stantibus*, the higher value must be disregarded.
4: Percentage depends upon the type of hereditament.

The contractor's method is often called a method of 'last resort' and this applies equally in rating valuation – it should only be adopted when the rental method is not possible.

3. *The Profits Method*

This is also a method which the valuer uses only when unable to adopt the rental method. It is often used in situations where there is some element of monopoly, either legal, due to statutory requirements (for example, public houses), or factual, due to circumstances (for example, cinemas). The monopoly element excludes use of the contractor's method since that method assumes the occupier has the choice of building his own premises. If a monopoly exists, this is not possible, either because of the licence that is legally required, or because the property is of a type where the numbers are necessarily limited.

After explaining the assumptions upon which the profits method is based, application of the method should be discussed.

The basic assumption is that the amount of profit that a property is capable of generating will be taken into consideration by the hypothetical tenant when deciding the rent he is prepared to pay. Since the required rent is that which a hypothetical tenant might pay, the average business person is considered to be running the particular business for a profit. An extremely good or bad business ability must be disregarded in favour of the average ability. Case law on this point may be found in *Watney Mann v. Langley (VO) 1966*. If actual trading accounts are available, they are often taken as the best guide, but these accounts will obviously indicate only the efficiency of the actual occupier. As they will be used in the computation of the occupier's income tax assessment, it should be borne in mind that the accounts will no doubt be presented in the most favourable form for that purpose. The Valuation Officer has a statutory authority to require any occupier to disclose details of the rent which they pay, but he is not empowered to require disclosure of business accounts. However, when an assessment is in dispute, the accounts will normally be disclosed if the Valuation Officer requests this.

A simple example should be provided to illustrate this method. Such an example is not provided here, since a detailed valuation and explanation is given in Question 5.7 to which the reader is referred.

A brief summary, such as that which follows, may be useful in concluding the answer. Of the three methods of valuation used to prepare rating assessments, the rental method is preferred. When it is not possible to use the rental method, either the contractor's or profits method will be used. Although there may be difficulties in applying the last two methods, it is usually possible, with care, to produce a satisfactory rating assessment on the required statutory basis.

Question 5.5

Consideration should first of all be given to the rent at which it is proposed No. 7 Little Street will be let, and the terms of letting. Are these facts material to the likely rating assessment? For rating purposes, the shop will be valued according to the tone of the list (for an explanation of this, see Question 5.3) and only if rental levels have fallen since the Valuation List came into force, will current rentals be used in preference to a tone of the list valuation.

A glance at the proposed rent of No. 7 Little Street (£32 000 per annum) compared with the letting of No. 11 Little Street in 1972 (£4 250 per annum), indicates that a tone of the list valuation will produce a lower gross value than current rental levels and should be the basis upon which the valuation is prepared.

Accordingly, the evidence used will be that of rents passing for similar properties during the year prior to the coming into force of the current Valuation List. At the time of writing, the current Valuation List is that which came into force on 1 April 1973, therefore evidence of rentals achieved during 1972 would have provided the basis of values in that Valuation List. Thus, the letting of No. 11 Little Street should assist in discovering the likely rating assessment of No. 7 Little Street.

As only one comparable transaction is quoted, the assumption should be made that No. 11 Little Street is representative of other nearby shops and, similarly, its rental value in 1972 correctly reflected rental levels in the street. This is an essential assumption, since it would otherwise be difficult to place reliance upon an isolated transaction. To emphasise the point, it might also be a stated assumption that the letting of No. 11 Little Street in 1972 was an open market transaction, with no connection between the parties, no premium was paid and the £4 250 per annum represented the full rental value, on the terms stated of the property at that time (see Question 5.4).

Before analysing the 1972 rental of No. 11 Little Street, it should be adjusted to conform with the statutory definition of gross value (see Question 5.2(a)), since this is the basis of assessment for shops in the 1973 Valuation List. Gross value assumes the landlord is responsible for all repairs and insurance, whereas in this case the landlord was responsible for external repairs only. The adjustment may be carried out as follows:

Let the rental value in terms of gross value = £y

If $7\frac{1}{2}$ per cent of £y represents a reasonable amount for internal repairs and insurance, then

$$y - 0.075y = £4\,250$$
$$0.925y = £4\,250$$
$$y = £4\,595$$

Rental value of No. 11 Little Street in 1972, in terms of gross value, say £4 600 per annum.

This rental value may now be analysed using the zoning method. Two zones of 6 m depth and a remainder will be adopted, halving back. Zone depths of 5 m may be used as an alternative, but, whichever is adopted, there must be consistency between zone depths used in analysis and those used in subsequent valuations.

Let the Zone A rental value per m^2	= £X	
Zone A = 6.5 m × 6 m × X	= $39X$	
Zone B = 6.5 m × 6 m × $\frac{1}{2}X$	= $19.5X$	
Remainder = 6.5 m × 6 m × $\frac{1}{4}X$	= $9.75X$	
Rear extension = 5.5 m × 4 m × $\frac{1}{6}X$	= $3.67X$	
[see note 1] [see note 2]		
Rental value	= $71.92X$	

$$71.92X = £4\,600$$
$$X = £63.96$$
$$\text{say £64 per m}^2$$

Notes

1: Although this rear extension was demolished in 1980, its value would have been reflected in the rent in 1972.

2: It is not clear from the question whether the rear extension formed part of the sales space, or was, for example, storage space. The assumption has been made here that it was not part of the sales space and although, strictly speaking, it should not be zoned, a proportion of the Zone A value per m^2 may be ascribed to it. Alternatively a 'spot' value may be adopted for this portion of the accommodation. In the absence of further information, it would be equally valid to assume that the extension was part of the sales area. This would consequently reduce the Zone A value per m^2, since the extension would then be included in the 'remainder'. Whichever assumption is adopted, it should be clearly stated. In practice, such a situation may not present a problem, since the valuer should, with some investigation (or perhaps from personal knowledge) be able to determine the actual use of this extension.

Having derived the Zone A rental value per m^2 of No. 11 Little Street, it may now be applied to the valuation of No. 7 Little Street.

Calculation of gross value of No. 7 Little Street

Zone A	= 6 m x 6 m x £64	= £2 304
Zone B	= 6 m x 6 m x £32	= £1 152
Remainder	= 6 m x 6 m x £16	= £ 576
	Gross value	£4 032
	say	£4 030

A gross value of £4 030 has a corresponding rateable value of £3 330, calculated as follows (see also Question 5.2(a)).

$$RV = GV - \text{statutory deduction}$$

In this case, the statutory deduction is:

$$£100 + \frac{1}{6} \text{ of GV in excess of £430}$$
$$= £100 + \frac{1}{6} \text{ of £3 600}$$
$$= £100 + £600$$
$$= £700$$

Rateable value = £4 030 − £700 = £3 330

The advice to your client is therefore that the rating assessment of No. 7 Little Street is likely to be gross value £4 030 rateable value £3 330. However, it would be wise to inspect the Valuation List and check the rating assessments of both No. 7 Little Street and other similar properties in the street. Should the actual gross value of No. 7 be appreciably higher than anticipated, or higher than the assessments of similar shops, then your client should be advised to submit a proposal to the Valuation Officer, seeking a reduction in the gross value of No. 7 Little Street.

Question 5.6

This question provides an example of the application of the rental method of valuation, both direct and indirect (see Question 5.4).

First of all, the information from the two adjoining shops should be analysed. Since the lettings were both in 1972, they will provide evidence regarding the levels of rental value in the street at that time. At the time of writing, the 1973 Valuation List is currently in force and therefore, according to the tone of the list provisions in *Section 20 of the General Rate Act 1967* (see Question 5.3), assessments in the Valuation List should be no higher than values prevailing during the year prior to the list coming into force — in this case, 1972.

An essential assumption is that the lettings of both shops 1 and 2 were arms length transactions to market value at the time, with no connection between landlord and tenant (see Question 5.4). In the calculations which follow, analysis and valuation will be carried out using the zoning method, adopting two 6 m zones and a remainder. It will also be assumed that the freehold rack rented capitalisation rate is 7 per cent.

Analysis of the letting of Shop 1, from the tenant's viewpoint

The valuer needs to cast his/her mind back to 1972, reconstructing the valuation surrounding the surrender of an existing lease in return for a new one at that time.

Tenant's 'present' interest, surrendered in 1972

Let full rental value on full repairing and insuring terms = £y

	£y
Full net rental value	
less Rent paid [see note 1]	£1 000 pa
Profit rent	£y − 1 000 pa
YP 3 years [see note 2] at 8 per cent and 3 per cent (tax 40 per cent) [see note 3]	1.615
Capital value	£1.615y − 1 615

Notes

1: Rent paid is on the same repairing terms as full rental value, therefore no adjustment is required.

2: Length of unexpired lease in 1972.

3: Traditional yield pattern. Remunerative yield 1 per cent above freehold yield has been adopted. An annual sinking fund is assumed available at 3 per cent, with tax payable at 40p in the £.

Tenant's 'proposed' interest in 1972

Full net rental value	£y
less Rent paid	£1 700 pa
Profit rent	£y − 1 700 pa
YP 7 years [see note 1] at 8 per cent and 3 per cent (tax 40 per cent)	3.361
Capital value	£3.361y − 5 714

Note

1: There is a rent review after 7 years therefore the proposed profit rent will only be available for that period of time.

The tenant should be no better and no worse off under the 'proposed' situation as he was under the 'present', therefore

$$\begin{aligned}
\text{Present interest} &= \text{Proposed interest} \\
£1.615y - 1\,615 &= £3.361y - 5\,714 \\
£4\,099 &= 1.746y \\
y &= £2\,348 \text{ pa}
\end{aligned}$$

Analysis of the letting of Shop 1 from the landlord's viewpoint

Landlord's 'present' interest in 1972

Rent received	£1 000 pa	
YP 3 years at 6 per cent [see note 1]	2.673	£2 673
Reversion to full net rental value	£y	
YP in perpetuity deferred 3 years at 7 per cent [see note 1]	11.661	£11.661y
Capital value		£2 673 + 11.661y

Note

1: Adopting traditional yield pattern. Term income is considered more secure than reversionary income, therefore it is assumed that a lower yield would be accepted during the first 3 years.

Landlord's 'proposed' interest in 1972

Rent received	£1 700 pa	
YP 7 years at 6 per cent	5.582	£9 489
Reversion to full net rental value	£y	
YP in perpetuity deferred 7 years at 7 per cent	8.896	£8.896y
Capital value	£9 489 + 8.896y	

The landlord should be no better and no worse off under the new situation than he was under the existing, therefore

$$\text{Present interest} = \text{Proposed interest}$$
$$£2\,673 + 11.661y = £9\,489 + 8.896y$$
$$2.765y = £6\,816$$
$$y = £2\,465 \text{ pa}$$

The full net rental value from the tenant's viewpoint is £2 348 pa and from the landlord's is £2 465 pa and a reasonable compromise between the two might be assumed to be £2 400 pa. However, the letting of Shop 1 is on full repairing and insuring terms — the tenant is responsible for the cost of repairs and insurance. This does not accord with the statutory definition of gross value (see Question 5.2) which assumes repairs and insurance to be the landlord's responsibility, therefore the £2 400 pa must now be converted on to this basis.

Assuming 10 per cent of the full net rental value to be a reasonable estimate of the cost of repairing and insuring the hereditament, then:

Full net rental value		£2 400 pa
plus Repairs and insurance	say	£ 250 pa
Rental value in terms of gross value		£2 650 pa

The rental value of £2 650 pa may now be analysed to determine the Zone A rental value per m^2.

Let $£X$ = Zone A rental value per m^2.

$$\begin{aligned}
\text{Zone A} &= 6\text{ m} \times 6\text{ m} \times X &&= 36X \\
\text{Zone B} &= 6\text{ m} \times 6\text{ m} \times \tfrac{1}{2}X &&= 18X \\
\text{Remainder} &= 6\text{ m} \times 4\text{ m} \times \tfrac{1}{4}X &&= 6X \\
\text{Rental value} & &&= 60X
\end{aligned}$$

$$60X = £2\,650$$
$$X = £44.17$$

Zone A rental value of Shop 1 in 1972 and in terms of gross value is £44.17 per m^2.

Analysis of the letting of Shop 2 from the tenant's viewpoint

In this case, an initial premium was paid and the annual equivalent of this capital sum must be added to the lease rent in order to determine the rental value of the shop in 1972. This also applies to the repairs costing £2 000, which the tenant agreed to carry out at the commencement of the lease.

$$\text{Annual equivalent of premium and repairs} = \frac{£7\,000}{\text{YP 14 years at 8 per cent and 3 per cent (tax 40 per cent)}} = \frac{£7\,000}{5.632}$$

$$= £1\,243 \text{ pa}$$

Full rental value on full repairing and insuring terms = £2 000 + £1 243 = £3 243 pa.

Alternatively, the tenant may be considered to have paid a capital sum of £7 000 in return for paying a lease rent less than the full rental value and, consequently, the enjoyment of a profit rent.

Let full rental value on full repairing and insuring terms = £y.

Full net rental value		£y
less Rent paid		£2 000 pa
Profit rent		£y – 2 000 pa
YP 14 years at 8 per cent and 3 per cent (tax 40 per cent)		5.632
Capital value		£5.632y – 11 264

In effect, the tenant has paid £7 000 for the value of this profit rent, thus

$$£7\,000 = £5.632y - 11\,264$$
$$£18\,264 = £5.632y$$
$$y = £3\,243 \text{ pa}$$

Full rental value on full repairing and insuring terms = £3 243 pa.

Analysis of the letting of Shop 2 from the landlord's viewpoint

Value of landlord's interest assuming the tenant pays an initial capital sum of £7 000:

Rent received	£2 000 pa	
YP 14 years at 6 per cent	9.295	£18 590
Reversion to full net rental value	£y	
YP in perpetuity deferred 14 years at 7 per cent	5.540	£5.540y
		£18 590 + 5.540y
plus Initial capital payment		£ 7 000
Capital value		£25 590 + 5.540y

This should be equal to the capital value if the tenant pays the full rental value, but no initial capital sum.

	£y
Full net rental value	
YP in perpetuity at 7 per cent	14.286
Capital value	£14.286y

$$£25\,590 + 5.540y = £14.286y$$
$$£25\,590 = £8.746y$$
$$y = £2\,926 \text{ pa}$$

The full net rental value of £2 926 pa seems a little low compared with the £3 243 pa calculated from the tenant's viewpoint and it will be assumed that the parties would agree at a rental value, on full repairing and insuring terms, of £3 100 pa. This must now be converted on to terms of gross value.

Assuming that 10 per cent of the full rental value represents the cost of repairing and insuring liabilities, then:

$$y - 0.1y = £3\,100 \text{ pa}$$
$$0.9y = £3\,100$$
$$y = £3\,444$$

The rental value of Shop 1, in 1972, in terms of gross value, say £3 450 pa.

This may now be analysed to find the Zone A value per m^2.

Let $£X$ = Zone A rental value per m^2.

Zone A	$= 8 \text{ m} \times 6 \text{ m} \times \quad X$	$= 48X$
Zone B	$= 8 \text{ m} \times 6 \text{ m} \times \frac{1}{2}X$	$= 24X$
Remainder	$= 8 \text{ m} \times 3 \text{ m} \times \frac{1}{4}X$	$= 6X$
Rental value		$= 78X$

$$78X = £3\,450 \text{ pa}$$
$$X = £44.23$$

Zone A rental value of Shop 2, in 1972, and in terms of gross value is £44.23 per m^2.

Attention should now be focused on the remaining shops in the street. Information in the question confirms that these are all the same size (5 m frontage by 18 m depth) and each has a gross value of £2 225 in the Valuation List.

The assessments of these shops should now be analysed to determine the Zone A gross value per m^2 that the £2 225 represents.

Let £X = Zone A gross value per m^2.

Zone A   = 5 m × 6 m × X = 30X
Zone B   = 5 m × 6 m × $\frac{1}{2}X$ = 15X
Remainder = 5 m × 6 m × $\frac{1}{4}X$ = 7.5X

Gross value in terms of Zone A   = 52.5X

$$52.5X = £2\,225$$
$$X = £42.38$$

Zone A gross value of the remaining shops in the street is £42.38 per m^2.

The subject property has a gross value of £3 000 and this must also be analysed to establish the Zone A value per m^2 that it represents.

Zone A   = 6 m × 6 m × X = 36X
Zone B   = 6 m × 6 m × $\frac{1}{2}X$ = 18X
Remainder = 6 m × 6 m × $\frac{1}{4}X$ = 9X

Gross value in terms of Zone A   = 63X

$$63X = £3\,000$$
$$X = £47.62$$

Zone A gross value of the subject property is £47.62 per m^2.

At this stage, it would be advisable to summarise the results obtained before proceeding further.

Shop 1	1972 Zone A rental value per m^2	£44.17
Shop 2	1972 Zone A rental value per m^2	£44.23
Remaining shops in the street	Zone A gross value per m^2	£42.38
Subject property	Zone A gross value per m^2	£47.62

It appears that the subject property is over-assessed both in relation to rental evidence in 1972 and the assessments of other similar properties in the Valuation List. However, before advising your client, it would be wise to inspect the property to ensure that it is truly comparable with the other properties. The question does not disclose the layout of the street or the relative positions of the various properties. For instance, is it possible that the subject property is in a more favourable trading position than the other properties? Does it have any other advantages not apparent from the present information available? If inspection reveals true comparability, then clearly your client should dispute the assessment.

Reliance should be placed, not on the 1972 transactions involving Shops 1 and 2, but on the assessments of the similar properties in the street, whose gross values represent a Zone A value of £42.38 per m^2.

Assessment of the subject property on this basis would result in a gross value of £2 670, as follows:

Zone A   = 6 m x 6 m x £42.38 = £1 526
Zone B   = 6 m x 6 m x £21.19 = £ 763
Remainder = 6 m x 6 m x £10.60 = £ 382
                                          _____
                       Gross value   £2 671
                                          _____

                       say   £2 670

Assessments of the similar properties in the Valuation List may be taken as the Valuation Officer's opinion of their value. Alternatively, if the assessments have ever been challenged, then their value in the List now will either be that agreed between the occupiers and the Valuation Officer, or the value decided by the Court. In either case, these assessments should be relied upon as evidence.

In *Lotus and Delta Ltd. v. Culverwell (VO) and Leicester City Council 1976*, it was held that the assessments of comparable properties were relevant, indicating the Valuation Officer's estimate of comparative values and could be referred to in disputes involving the assessments of similar hereditaments. In the light of the above, your client should be advised to serve a proposal on the Valuation Officer. This should specify that the gross value of the subject hereditament be reduced from £3 000 to £2 670 on the grounds that it is incorrect and excessive compared with the assessments of similar properties in the Valuation List.

Question 5.7

Before commencing the valuation, a few general points might be made:

(i) The valuation will be carried out using the profits method.
(ii) Extracts from the accounts have been averaged over the last 3 years. This is the usual procedure so long as they represent trade that can be reasonably maintained in the future. In the absence of information to the contrary, it will be assumed that this is the case.
(iii) Since the statutory definition of gross value assumes that the landlord repairs and maintains the property, any item in the accounts relating to repairs or insurance of the buildings, are excluded from the calculation.
(iv) The aim of the valuation is to determine the rating assessment, which is based on the rent that a hypothetical tenant would pay for the premises, therefore amounts shown in the accounts for rent and rates will also be excluded from the calculation.
(v) In the valuation, an average business person is assumed, therefore repayment of

interest on loans is excluded. High bank charges due to an overdraft would be similarly treated.

(vi) Occupier's drawings are disregarded in the working expenses, since these are accounted for in the tenant's share.

The profits method of valuation is also dealt with in Question 5.4.

Valuation

Receipts: Restaurant		£ 47 500
Bars		£ 71 000
Letting of rooms		£ 32 500
Total receipts		£151 000
less Purchases		£ 55 250
Gross profit		£ 95 750

Less working expenses:		
Wages, salaries and National Insurance	£ 36 000	
Insurance (contents and third party)	£ 240	
Laundry and cleaning	£ 3 200	
Advertising, stationery etc.	£ 890	
Postage and telephone	£ 545	
Repairs to furniture and fittings	£ 2 800	
Accountant's fees	£ 375	
Bad debts	£ 175	
Lighting and heating	£ 3 240	
Annual sinking fund to replace tenant's chattels	£ 1 025	£48 490
Net profit		£47 260

Less Interest on tenant's capital: [see note 1]		
Furniture and equipment	£ 77 000	
Consumable stock	£ 12 000	
Cash	£ 6 000	
	£ 95 000	
at say 10 per cent		£ 9 500
Divisible balance [see note 2]		£37 760
Tenant's share 40 per cent		£15 104
Rent and rates		£22 656

Notes

1: The tenant is assumed to provide these items at his own expense. He is therefore also assumed to be entitled to interest, on the capital he has provided, from the business.

2: This is an amount sufficient to remunerate the tenant for running the business. The percentage adopted will depend upon the circumstances in a particular case, including the type of business and the risks involved. As an alternative, the figure may be expressed as a percentage of turnover. In the example, £15 100 represents 10 per cent of turnover.

The amount available for payment of rent and rates is £22 656 and it now remains to separate these two items.

A formula has to be used for this purpose:

$$\text{Rent} + \text{rates} = \text{gross value} + (\text{net annual value} \times \text{rate in the £})$$
$$\text{GV} = \text{NAV} + \text{statutory deductions}$$

therefore

$$\text{rent} + \text{rates} = (\text{NAV} + \text{statutory deductions}) + (\text{NAV} \times \text{rate in the £}).$$

The table of statutory deductions is provided and explained in the answer in Question 5.2, but in order to solve the present problem, it is necessary to express these deductions in terms of net annual value, rather than the usual gross value.

These are as follows:

Gross value	*Statutory deductions in terms of NAV*
Not exceeding £65	$\frac{9}{11}$ NAV
Exceeding £65 but not exceeding £128	$13.57 + \frac{3}{7}$ NAV
Exceeding £128 but not exceeding £330	$\frac{\text{NAV}}{5} + 32$
Exceeding £330 but not exceeding £430	$\frac{\text{NAV} + 70}{4}$
Exceeding £430	$\frac{\text{NAV}}{5} + 34$

Thus, in the example, where the rate in the £ is 292p:

$$£22\,656 = \left(\text{NAV} + \frac{\text{NAV}}{5} + 34\right) + (\text{NAV} \times £2.92)$$

$$£22\,656 = (1.2\,\text{NAV} + 34) + 2.92\,\text{NAV}$$

$$£22\,656 = 4.12\,\text{NAV} + 34$$

$$£22\,622 = 4.12\,\text{NAV}$$

$$\text{NAV} = £5\,491$$

In this instance, gross value is required.

$$\text{gross value} = \text{NAV} + \text{statutory deductions}$$

$$= 5\,491 + \frac{5\,491}{5} + 34$$

$$= £6\,623$$

The valuation prepared, as required, for rating purposes, produces an assessment of gross value £6 623. This could be rounded off to £6 625, giving a corresponding rateable value of £5 493 (£6 625 minus statutory deduction of £100 + 16.66 per cent of £6 625 − £430).

Your client should therefore be advised to submit a proposal to the Valuation Officer, specifying that the rating assessment of his hotel should be reduced from gross value £7 250 rateable value £6 014 to gross value £6 625 rateable value £5 493. Grounds for reducing the assessment must be given in the proposal and, in this case, it is suggested that it would be sufficient to state that the present assessment is incorrect and excessive.

6 Compulsory Purchase Compensation

The questions in this chapter deal with most of the situations, which may be encountered in compulsory purchase claims, namely:

(i) Land taken in accordance with *Rules 1 to 5, Land Compensation Act 1961* and the planning assumptions contained in *Sections 14 to 16 of the Act.*

(ii) Disturbance and other costs to the person in accordance with *Rule 6, Land Compensation Act 1961* and injurious affection to land retained where land is taken.

(iii) Injurious affection where no land is taken, as provided in the *Land Compensation Act 1973.*

(iv) The acquisition of unfit housing in accordance with the *Housing Act 1985 and the Land Compensation Act 1973.*

In some instances, the questions require assumptions to be made and, where this occurs, these are fully explained.

In addition to claim figures calculated, professional fees will be paid by the acquiring authority.

COMPULSORY PURCHASE COMPENSATION – QUESTIONS

6.1. The freehold site of a factory (original area 1000 m^2 net), built 100 years ago and demolished 5 years ago, is to be compulsorily acquired for public open space, for which purpose it is allocated in the Development Plan. The District Council has decided to acquire the site because the owners of adjoining residential property demand its use as open space.

The site has been occupied for the past 3 years on a 7 year lease for the storage of timber at a rent of £1 000 per annum. The full rental value is considered to be £2 000 per annum. The rateable value is £300.

Prepare a claim for the freeholder of the site, making necessary assumptions to show that all possibilities have been considered.

6.2. (a) Your client is the owner–occupier of a large detached house in 0.5 hectares of ground situated in open country adjoining a disused airfield. Central Government is proposing to purchase the airfield as a dumping ground for nuclear waste, which will entail lorries directly passing your client's property and the use of searchlights and patrols.

There is no intention to acquire any of your client's land.

Advise your client regarding his compensation rights.

(b) Under *Sections 14 to 16 of the Land Compensation Act 1961*, there are certain planning assumptions which a valuer should consider when assessing compensation.

Discuss these assumptions and how *Rule 3, Section 5 of the Land Compensation Act 1961* might have some effect upon them.

6.3.

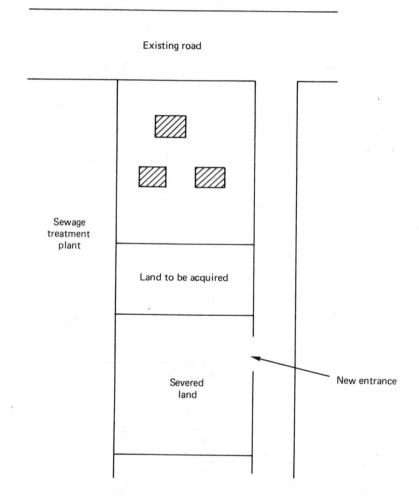

Existing road

Sewage treatment plant

Land to be acquired

Severed land

New entrance

A small farm occupies a 0.75 hectare site.

The local authority wish to acquire 250 m² of land to provide a new access road to a sewage treatment plant, which is adjacent to the farm.

This will create a severed portion of land having an area of 0.25 hectares.

Prepare a compensation claim on behalf of the freehold owner of the farm.

6.4. Your client, the freehold occupier of a factory having 1000 m² of net usable space, has recently been served with a Notice to Treat.

His gross annual profits over the past five years have been:

£17 000, £17 500, £17 500, £18 000, £18 500.

Prepare a compensation claim on the assumption:

(i) that your client is unable to relocate elsewhere,

(ii) that your client has found alternative premises nearby, but the move means that his business will be closed down completely for a period of 6 weeks and, in consequence, he will lose permanently 10 per cent of his customers. However, it is anticipated that the business will be operating at its former level of profit within two years of the move.

6.5. A is the freehold owner of a factory on the outskirts of a provincial city, having net usable floor area of 5000 m² on three floors, the site having an area of 3000 m². The premises are let to B on a 14 year full repairing and insuring lease having 5 years unexpired at a rent of £40 000 per annum. The current net rental value is considered to be £50 000 per annum.

Notice to Treat has been served on both A and B by an acquiring authority as the site is required as playing fields to form part of a leisure complex.

The acquiring authority has offered B a factory unit on its new industrial estate, also 5000 m² net, at a rent, on full repairing and insuring terms, of £25 per m².

Details from the company's most recent profit and loss account and balance sheet reveal the following:

Gross receipts from sales	£600 000
Purchase of materials	£200 000
Wages	£120 000
Directors' emoluments	£ 15 000
Other working expenses including rent	£ 80 000
Stock in hand	£ 80 000
Plant and Machinery (cost *less* depreciation)	£120 000
Capital employed in business	£ 60 000

Prepare compensation claims for A and B, taking all possibilities into account.

6.6. A 250 year old church, in a poor state of repair, is situated in a mixed office and shopping area. The provisions of the local plan provide that part of the area (including the church) is to be redeveloped for old people's flats and neighbourhood shopping centre.

The church site has 18 m frontage and 30 m depth, and the local authority wish to compulsorily acquire the site having just served a Notice to Treat.

The church has a declining congregation of about 30 persons and services take place only once a month. The trustees have been concerned about this and are keen to have a new church in another part of the town where they believe a revival will take place because of a new housing development. They have found a suitable freehold site and obtained planning permission. The freehold owner of the site requires £30 000 and this is accepted as being the market value.

Advise the trustees as to how compensation may be assessed, and calculate the likely compensation.

6.7. A is the freehold owner of 4 adjoining terraced houses, which he purchased 20 years ago. He occupies one himself and the other three are each let on regulated rents, the current fair rent being £30 per week exclusive. The rateable value of each house is £110, and each has a vacant possession value of £20 000.

One of the tenanted houses is on a site zoned as being suitable as a small shop site worth £10 000.

The houses are 'unfit' and the local authority intend to acquire the site under the *Housing Act 1985*.

Prepare a compensation claim on behalf of the freeholder.

COMPULSORY PURCHASE COMPENSATION – SUGGESTED ANSWERS

Question 6.1

In answering this question, it is necessary to prepare three valuations to establish which gives the highest figure of compensation.

(i) The first approach may be to assume the rebuilding of the factory in 4 years' time. This would be permissible under *Section 15(3), Land Compensation Act 1961* which assumes that *Part I, Eighth Schedule, Town and Country Planning Act 1971 (subject to 18th Schedule)* may apply – the rebuilding of the factory plus 10 per cent tolerance. The value of the land on this basis would need to be deferred by the 4 year unexpired period of the lease, so that the claim might be:

Rent reserved	£ 1 000 pa	
YP 4 years at 10 per cent [see note 1]	3.17	£ 3 170
Reversion to development value of land [see note 2]	£56 950	
PV of £1 in 4 years at 9 per cent [see note 3]	0.708	£40 320
		£43 490

Less

Compensation to tenant [see note 4]			
3 x Rateable value		£ 900	
PV of £1 in 4 years at 10 per cent		0.683	£ 615
	Compensation claim		£42 875

Notes

1: It is considered that an initial yield of 10 per cent would be appropriate for a letting of timber storage.

2: Because no comparables are available and the site area is not given, a residual method may be used to assess development value. Rentals of £45 per m^2 and building costs of £300 per m^2 have been used:

1000 m^2 net + 10 per cent tolerance =		
1100 m^2 x £45		= £ 49 500 pa
YP in perpetuity at 8 per cent		12.5
Gross Development Value		£618 750

Less

(i)	Demolition	say	£ 5 000	
(ii)	Building costs 1 100 m^2 + 10 per cent =			
	1210 m^2 gross x £300		= £363 000	
(iii)	Contingencies 10 per cent of			
	demolition and building costs		= £ 36 800	
(iv)	Quantity surveyor's and architect's			
	fees 10 per cent of (i), (ii) and (iii)		= £ 40 480	
(v)	Interest charges say 12 per cent on			
	half (i)–(iv) for 1 year	say	£ 26 700	
(vi)	Estate agency and legal fees 3 per cent			
	of Gross Development Value		= £ 18 560	
(vii)	Developers profit 10 per cent of			
	Gross Development Value		= £ 61 875	£552 415
	Residue available for land, acquisition			
	and finance costs			£ 66 335

Assume land value = X, and acquisition costs are 4 per cent, then residue = $1.04X$ + $0.12(1.04X) = 1.1648X$

$$\text{Value of land} = £\ \frac{66\,335}{1.1648} = £56\,950$$

3: A discount rate of 9 per cent is appropriate for development.

4: The compensation payable to the tenant on expiration of his lease will be 3 × Rateable Value in accordance with the *Landlord and Tenant Act 1954.*

(ii) The second approach may be on the basis of existing use by the investment method:

Rent reserved	£1 000 pa	
YP 4 years at 10 per cent	3.17	£ 3 170
Reversion to full net rental value	£2 000 pa	
YP in perpetuity deferred 4 years at 10 per cent	6.83	£13 660
Compensation claim		£16 830

(iii) The third possibility may be to apply a *Section 17 certificate* procedure under the *Land Compensation Act 1961,* i.e. a certificate of appropriate alternative development, which may allow residential development. This would give rise to a claim based on the area of the land (not given) by a capital value per acre or hectare.

Dependent on the figures used, approaches (i) and (iii) will give higher claims than approach (ii).

Question 6.2

(a) Any claim in this situation will be in relation to compensation where no land is taken.

For a usage as sensitive as nuclear waste dumping, there will be an Act of Parliament which should authorise the works. However, there may be no specific compensation provisions.

Reference may be made to *Section 10, Compulsory Purchase Act 1965,* the effect being clarified by the four rules laid down in *Metropolitan Board of Works v. McCarthy (1874).* These deal with injurious affection caused by the lawful execution of the works which would have been actionable but for the statutory powers to proceed with the works. Rule (iv) states "the damage must arise from the execution of the works and not the subsequent use." In this case, it is more likely that depreciation will be due to the eventual use of the site, not its preparation, so that this section is ineffective in this situation.

The provisions of *Part I, Land Compensation Act 1973* will apply in this case. This Act greatly increased the scope for injurious affection compensation where no land is taken; it is payable for depreciation in the value of land by the use of certain public works caused by physical factors such as noise, vibration, smell, fumes, smoke and artificial lighting and the discharge onto land of any solid or liquid substance. The claim will be limited to physical factors and not items such as loss of view or privacy. In this case, the depreciation caused by noise and artificial lighting may be claimable on the basis of a 'before' and 'after' valuation:

Open market value assuming works are in existence but without 'physical factors'	£100 000
Less Open market value with the effect of 'physical factors'	£ 80 000
Compensation claim	£ 20 000

(b) *Section 14(1) of the Land Compensation Act 1961* provides for the taking into account of the *six Assumptions* as to planning permission set out in *Sections 15 and 16*, when estimating open market value.

Section 14(2) allows for any planning permission actually in force at the date to be taken into account.

Section 14(3) permits the payment of 'hope value' if the evidence is that the market would allow it. While this is stated, it may be difficult to establish in practice and a value on this basis may be heavily discounted.

Section 15(1) considers permission as would permit development in accordance with the proposals of the acquiring authority. It is this section which may conflict with *Rule 3, Land Compensation Act 1961* which states "The special suitability or adaptability of the land for any purpose shall not be taken into account if that purpose is a purpose to which it could be applied only in pursuance of statutory powers, or for which there is no market apart from the special needs of a particular purchaser of the requirements of any authority possessing compulsory purchase powers."

This rule suggests that value generated by a very specialised use which could only be initiated by the acquiring authority (such as a crematorium) should be disregarded.

Where the land is being acquired for a more common use such as housing, the private sector could effect this use and it may be taken into account. Further conflict may arise if the *'Pointe Gourde' principle* is applied — "Compensation cannot include an increase in value which is entirely due to the scheme underlying the acquisition" *(Pointe Gourde Quarrying and Transport Company Limited v. Sub-Intendent of Crown Lands 1947)*.

Section 15(3) and (4) allows permission to be assumed for both *Part 1 and Part 2, Eighth Schedule, Town and Country Planning Act 1971*, unless compensation under Part 2 has become payable. *Section 16(2)* allows permission to be assumed for development for which planning permission might reasonably be expected to be granted for zoned use in a Development Plan.

Section 16(4) and (5) states that in Comprehensive Development Areas or Action Areas the planned range of users in the area has to be considered and then there must be determined that which might have been appropriate had there been no such defined area and the Development Plan had not contained any proposals for the area.

It must be further assumed that no development or redevelopment had taken place in the area in accordance with the Plan and none of the land was proposed to be acquired for public purposes.

The philosophy of this Section is to reveal what consents are likely in a 'no-scheme' situation.

Question 6.3

This claim will entail a valuation of the land taken and injurious affection in relation to the land severed by the acquisition plus essential works.

Claim

Land taken	
250 m^2 (0.025 ha) at £7 500 per ha [see note 1]	= £ 188
Land subject to injurious affection 0.25 ha at	
£3 750 per ha [see note 2]	= £ 937
Essential works	
Formation of new entrance to severed land [see note 3]	say £ 250
Compensation claim	£1 375

Notes

1: It is considered that £7 500 per ha. would be a reasonable capital value for land attached to a small farm of this type.

2: It has been assumed that the severed land has reduced in value per hectare by 50 per cent of the original. Because it is no longer attached to the area containing farm buildings, access is inconvenient, assuming a new entrance is formed from the side road. The farmer will suffer more inconvenience which is reflected in the diminution of value.

3: A gated entrance will be required to the severed land and alterations to fencing.

Question 6.4

(i) *Compensation for freehold interest*
    [see note 1]
    Rental value = 1000 m^2 × £25   = £25 000 pa
    [see note 2]
    YP in perpetuity at 12 per cent   8.334
    [see note 3]   £208 350

Compensation for extinguishment of business [see note 4]			
Average gross profit		£17 700	
Less			
Interest on capital:			
Stock	say	£10 000	
Fixtures and fittings	say	£ 5 000	
Capital	say	£10 000	
		£25 000	
12 per cent on £25 000 =		£ 3 000	
Average net profits		£14 700	
YP [see note 5]	say	$1\frac{1}{2}$	£22 050
Enforced sale of fixtures and fittings and stock [see note 6] — loss say 25 per cent on £15 000			£ 3 750
Redundancy payments [see note 7]			say £ 5 000
Incidentals			say £ 1 000
Compensation claim			£240 150

Notes
1: The compensation for the freehold factory will be claimable under *Rule 2, Section 5, Land Compensation Act 1961*. There is no suggestion in this question of alternative use or planning assumptions.
2: The rental of £25 per m^2 has been assumed as reasonable for a factory, where little information is given.
3: The initial yield of 12 per cent is considered to be appropriate for this type of property.
4: Compensation for extinguishment of the business is claimable under *Rule 6, Section 5, Land Compensation Act 1961.*
5: This disturbance claim requires the calculation of average gross profit from the last five years' figures. In this case, these have been fairly constant.
   The capital 'tied' into the business has been assumed to be £25 000, and it is

considered that 12 per cent is a reasonable interest rate. Having obtained average net profits, a YP is applied to this, which reflects the economic activity of the business and its potential growth. Because of the steady nature of this business, $1\frac{1}{2}$ YP is applied.

6: It will be necessary to sell fixtures, fittings and stock within a certain time period so that it is anticipated that 25 per cent loss on enforced sale might be suffered.

7: Because of insufficient information, an amount has been assumed for redundancy payments. This would be based on age of employees, whether male or female and years of service. The payment is expressed as a number of week's wages.

(ii) *Compensation for freehold interest as before*			£208 350
Temporary loss of profits [see note 1]			
— Six weeks' closure based on £14 700 per annum	say £1 700		
Two years' loss in profits say 10 per cent of £14 700 per annum		£2 940	£ 4 640
Cost of finding alternative accommodation, removals and expenses [see note 2]		say £ 5 000	
			£217 990
	Compensation claim	say £218 000	

Notes

1: This type of claim is often difficult to quantify and there are few Lands Tribunal decisions to provide guidance. It is for the claimant to substantiate his loss. In this instance, the loss on six week's closure is calculable, but the permanent loss of 10 per cent of existing customers is problematic. The calculation has assumed 10 per cent lost profit over two years; this may be too simplistic!

It has been assumed that 'former level of profit' means the level which would have been achieved in two years' time in the former premises, so that there is no claim for loss of further profits after two years.

2: £5 000 has been used to cover the cost of finding alternative accommodation, removal costs, changing letterheads and signs on vehicles, adapting fixtures and fittings and telephones.

Although perhaps not relevant to this answer, care must be taken to ensure that double overheads are identified, i.e. those overheads that are incurred at both old and new premises.

Question 6.5

The compensation for the freeholder may be calculated as follows:

A's claim

Rent reserved	£40 000 pa	
YP 5 years at 8 per cent [see note 1]	3.993	
	———	£159 720
Reversion to net rental value	£50 000 pa	
YP in perpetuity deferred 5 years at 8 per cent	8.507	£425 350
	Compensation claim	£585 070

Note

1: An initial yield of 8 per cent is considered suitable for both term and reversion.

The planning assumptions, *Sections 14 to 16, Land Compensation Act 1961*, may be considered although playing field value might be less than existing use. If the leisure complex could not be built without the playing fields, the philosophy of *Stokes v. Cambridge Corporation (1961)* might apply, i.e. an owner holding the 'key' to a development might benefit from some proportion of the development value. It is assumed here that this is not appropriate.

The compensation payable to the lessee will be the value of his lease in accordance with *Rule 2, Land Compensation Act 1961* and disturbance payments in accordance with *Rule 6, Land Compensation Act 1961*.

B's Claim

Value of lease

Full rental value	£50 000 pa	
Less Rent paid	£40 000 pa	
Profit rent	£10 000 pa	
YP 5 years at 9 per cent and 3 per cent		
(tax 40 per cent)	2.476	£24 760

Disturbance

This claim will be dependent upon whether the tenant's business is totally extinguished or whether he relocates elsewhere (in the local authority unit or he finds alternative accommodation himself).

He currently occupies premises having a rental value of £10 per m², but is being offered alternative accommodation at £25 per m². B may prove that the company

Advanced Valuation

could not afford to pay this rent and that he could not re-locate elsewhere. If the acquiring authority accepted total extinguishment, the claim would be:

Gross receipts		£600 000	
Less Purchase of materials	£200 000		
Wages	£120 000		
Directors' emoluments [see note 1]	£ 25 000		
Other working expenses including rent	£ 80 000	£425 000	
Net profit		£175 000	
Less			
Profit rent [see note 2]	£10 000		
Interest on stock, plant and machinery and capital at 12 per cent [see note 3] = 0.12 × £260 000	£31 200	£ 41 200	
Adjusted net profit		£133 800	
YP [see note 4] say		2	£267 600
Redundancy payments [see note 5] say			£ 10 000
Enforced sale of stock, plant and machinery [see note 6] say 30 per cent of	£180 000	£ 54 000	
Incidentals say		£ 5 000	
			£336 600
Thus, compensation claim is £24 760 + £336 600 =			£361 360

Notes

1: It is considered that £25 000 is a more realistic figure for directors' emoluments than £15 000 provided in the accounts. According to *Shulman (J) (Tailors) v. Greater London Council 1966*, a proper amount should appear.

2: Profit rent is deducted, as it has already been allowed for in the valuation of the lease.

3: When the business comes to an end, the capital employed in the business of £260 000 is released and might be invested at a rate of interest appropriate at the time, say 12 per cent. (The case of *Handley v. London Borough of Greenwich 1970* adopted 10 per cent.)

4: The adjusted net profit has been based on this year's accounts although an average of the last 3 years' profits is preferable. The YP is based on the state of the business, and whether profits are increasing or not. A YP of 2 may indicate a fairly thriving business.

5: Redundancy payments would be paid in accordance with statutory limits.

6: Because stock, plant and machinery may have to be sold in a short period of time, the price attained will be less than market value so that the loss is claimable.

OR

If it is assumed that B would be re-located in the local authority's premises or he found alternative premises himself, then his compensation claim would include value of the lease, costs of alternative premises, and permanent and temporary loss of profits. The rent to be paid for the alternative premises is immaterial to the claim as it is assumed that the lessee obtains premises of a quality reflected by the rent.

B's claim

Value of the lease as before		£24 760
Disturbance		
Costs attached to the alternative premises [see note 1]		
Cost of removals	say £5 000	
Publicity costs	say £1 500	
Directors' fees searching for new premises	say £3 000	
Abortive professional fees	say £2 000	
Double overheads	say £5 000	£16 500

Note

1: The cost of removals would include dismantling, adapting and re-installing plant, machinery, fixtures and fittings and also loss on sale of those items which could not be re-used.

Publicity costs might include notification of new address, reprinting stationery, altering signs on motor vehicles and telephone removals.

Directors' fees searching for new premises might include travelling expenses and reasonable loss of earnings while searching for premises.

Abortive professional fees would be paid if reasonable — probably an area of contention!

Claims for double overheads will arise if it is necessary to operate two premises for an overlap period. This might include double rent, rates, heating, lighting and telephone.

Permanent loss of profit

It is assumed that when the business is in its new location, the profit level will reach 75 per cent of its existing level when full production is achieved.

Value to the claimant of goodwill was calculated earlier to be £267 600, so that 25 per cent of this gives a claim of £66 900.

Temporary loss of profit

Assume a run-down period of 6 months prior to the move and that this will affect the profits by 25 per cent. The actual move will take a month and the run-up to 75 per cent of previous performance takes 6 months.

Run-down

Existing adjusted net profits	£133 800	
6 month's loss at 25 per cent	0.125	£16 725

Month's loss

New adjusted net profit =		
£133 800 − 25 per cent	£100 350	
1 month's loss	0.0834	£ 8 370

Run-up to full production

Existing adjusted net profits	£133 800	
6 month's loss at 25 per cent	0.125	£16 725
		£41 820

Thus, claim would be £24 760 + £16 500 + £66 900 + £41 820 = £149 980.

The local authority would obviously prefer B's claim for relocating elsewhere rather than total extinguishment, although this may seem a high price to pay for playing fields.

Question 6.6

The trustees may proceed under *Section 5 of the Land Compensation Act 1961* to assess compensation in accordance with *Rule 2* or *Rule 5*.

Rule 2 is the 'open market value', taking into account the planning assumptions as specified in *Section 15(1)* — "that which would permit development in accordance with the proposals of the acquiring authority" and *Section 16(2)* — "development for which planning permission might be reasonably expected to be granted for the zoned use in the development plan." In this case, this might suggest shops.

So that value of site for shops:

18 m × 30 m × £200/m^2 [see note 1]	= £108 000
Less Cost of demolition	= £ 10 000
Site value	£ 98 000

The trustees might consider an alternative use value such as storage, assuming that planning permission could be obtained.

So that value for storage use:

Net rental value = 18 m × 30 m × £15/m^2	= £ 8 100 pa	
[see note 2]		
YP in perpetuity at 12 per cent [see note 3]	= 8.334	
Capital value	say £67 500	

Notes

1: The value of £200/m^2 for commercial development is considered to be appropriate.

2: It is considered that £15/m^2 is a suitable rent for old church premises used for storage.

3: An initial yield of 12 per cent would be acceptable for this type of property.

The trustees would then be advised to prepare a claim under *Rule 5* — "Where land is, and but for the compulsory acquisition would continue to be, devoted to a purpose of such a nature that there is no general demand for that purpose, the compensation may, if the Lands Tribunal is satisfied that reinstatement in some other place is bona fide intended, be assessed on the basis of the reasonable cost of equivalent reinstatement."

It seems to be generally accepted that there is no general demand for church premises. The trustees have indicated in these circumstances that they do intend to rebuild elsewhere, having obtained planning permission on an alternative site. The practicality of relocating with a small congregation may be questioned, but the trustees are supported by *Zoar Independent Church Trustees v. Rochester Corporation 1974* which involved a church with twelve members.

A *Rule 5* claim may, thus, be prepared:

Purchase of alternative site	£ 30 000
Legal and other fees on above say 4 per cent	£ 1 200
Cost of equivalent building — 18 m × 30 m × £400/m^2	£216 000
[see note 1]	
Claim	£247 200

Note

1: For simplicity, the £400/m^2 is inclusive of building costs, contingencies, fees and financing. A more detailed claim might itemize these separately.

The trustees are advised to submit a claim in the sum of £247 200.

Question 6.7

This compensation claim will consist of three situations:

(i) The owner-occupied house will have a claim including site value, owner-occupiers supplement, a disturbance allowance and a Home Loss payment.
i.e.

Site value			£10 000
Add Owner–occupiers supplement			
Vacant possession value		£20 000	
Less Site value		£10 000	
			£10 000
Disturbance claim	say		£ 500
Home loss payment 10 × RV [see note 1]			£ 1 100
			£21 600

Note

1: The freeholder is entitled to a Home Loss payment in accordance with *Section 30(1) (b) Land Compensation Act 1973*. (From 16 January 1989, the Home Loss Payment is based on 10 × Rateable value.)

(ii) The tenanted house having a value for a small shop site of £10 000.

i.e. Site value	£10 000
plus Well-maintained allowance 14 × RV [see note 1]	= £ 1 540
	£11 540

Note

1: Well-maintained allowance of 14 × RV is in accordance with *Section 586 and Schedule 23, Housing Act 1985*. It is assumed that the freeholder is responsible for all repairs and he has maintained the property satisfactorily.

OR

Ceiling value – investment value	
Gross income 52 weeks × £30	= £ 1 560 pa
Less Repairs, insurance and management say 20 per cent	
of income	say £ 300 pa
Net income	£ 1 260 pa
YP in perpetuity at 12 per cent	8.334
	£10 500

The compensation is restricted to the ceiling value of £10 500.

(iii) The other 2 tenanted houses. While these sites are not specifically zoned for small shop use, the claim might follow the same pattern as (ii) above, so the compensation for each would be £10 500.

So total claim is

Owner-occupier house	£20 850
Three tenanted houses	£31 500
Compensation claim	£52 350

The tenants of the houses do not have compensatable interests, and are not entitled to a notice to treat. They will be rehoused by the local authority and may receive disturbance payments under *Section 37, Land Compensation Act 1973.*

7 Planning Compensation

Compensation in relation to adverse planning decisions is predominantly governed by provisions in the *Town and Country Planning Act 1971*.

Students may encounter the following situations in examination questions:

1. Refusal of planning permission in relation to new development. The entitlement to compensation in this case is limited to a maximum payment equal to the *unexpended balance of established development value (UXB)*. The general rule is 'No UXB — No Compensation'. Question 7.2 requires an explanation as to how the UXB has developed from the original right for owners of legal estates to establish claims under *Part VI, Town and Country Planning Act 1947*.

It should be remembered that owners aggrieved by an adverse decision might serve a purchase notice, the onus being upon them to prove that the property, is "incapable of reasonable beneficial use."

2. Adverse decisions in relation to extensions to buildings which are 'permitted development' in accordance with *Part II, Eighth Schedule, Town and Country Planning Act 1971*. The area and/or cubic capacity of the extension in relation to the area/cubic capacity of the original building should not exceed the limits laid down in General Development Orders currently prevailing; at the time of writing basically 10 per cent.

This development of a marginal nature, if refused, will give rise to claims for compensation in accordance with *Section 169 of the 1971 Act*.

The level of compensation will be the difference between value of the property with the extension and its existing use value.

Question 7.6 requires the preparation of a compensation claim in this type of situation.

It is important to differentiate between *Part I and Part II of the Eighth Schedule*, as Part I situations do not have compensatable rights. This includes the rebuilding or integral alteration of existing buildings and the conversion of a single dwelling house into two or more separate units. A refusal to allow Part I development might lead to the service of a purchase notice by the aggrieved owner (as in situation 1).

3. Revocation or modification of an existing planning permission. *Section 164 of the 1971 Act* deals with compensation in these cases, and it may be dealt with under two heads:

(a) Loss in land value due to the decision. This will be the difference between the value with the benefit of planning permission and the value with the effect of revocation or modification.

(b) Abortive expenditure, and loss or damage directly attributable to the revocation or modification. This may include professional fees and liquidated damages to the contractor if he is prevented from carrying out building works. The revocation/modification order is not retrospective so that work already carried out is not affected. This might lead to a claim for demolition costs, if there are partially completed buildings.

Questions 7.1, 7.4 and 7.5 deal with revocation situations and require residual valuations in 7.1 and 7.4 and the use of a discounted cash flow in 7.5.

These situations give the opportunity to test general areas of valuation such as analysis and other legislative valuation such as landlord and tenant, and leasehold enfranchisement.

4. Discontinuance of an established use. *Section 170 of the 1971 Act* deals with compensation in these cases, and it may be dealt with under three heads:

(a) Loss in land value. This may be a 'before' and 'after' valuation, i.e. existing use value less the value with the effect of discontinuance.

(b) Disturbance. There may be partial or total extinguishment of a business and the amount claimed may follow the pattern established for compulsory purchase claims.

(c) Expenses incurred. This might include such items as loss on enforced sale of stock, fixtures and fittings and redundancy payments.

Question 7.3(a) deals with this situation. Discontinuance orders should not be confused with enforcement notices, which seek to bring to an end unauthorised activities in relation to land and/or buildings.

5. Adverse decisions in relation to more specialised situations such as listed buildings and tree preservation orders.

Where a building is 'listed' by the Secretary of State and its owner wishes to alter or extend it, he has to obtain listed building consent. If this is refused, he has compensation rights under *Sections 171–173 of the 1971 Act*.

The claim will be the difference between value of the property with the alteration or extension and its existing use value.

There are also compensation provisions in relation to building preservation notices and ancient monuments.

If a building preservation notice is served without the listing of the building, *Section 173* provides for the payment of compensation for loss or damage directly attributable to the effect of the notice.

Similar provisions apply to historic and ancient monuments protected under the *Historic Buildings and Ancient Monuments Act 1953*.

Tree preservation orders may present problems to prospective developers by seriously restricting the use of development land. Application for consent to fell such trees may be made and, if this is refused, compensation under *Section 174 of the 1971 Act* may be payable. This will be the difference between the value of development land (with trees removed) and the value with the trees still standing. The alternative is the service of a purchase notice as before.

Question **7.7** deals with compensation claims in relation to listed buildings and tree preservation orders.

In all claims, the claimant will be entitled to receive professional fees incurred in relation to the preparation and settlement of his claim.

PLANNING COMPENSATION − QUESTIONS

7.1. Your client is the head lessee of a site with a factory upon it, the lease having 50 years unexpired at a fixed ground rent of £200 per annum. The site is sublet on a full repairing and insuring lease with 3 years unexpired at a fixed rent of £2 000 per annum. The current net rack rental value is considered to be £3 000 per annum.

Your client has obtained planning permission to provide 1200 m^2 net of offices on the site, which will be let on 5 year full repairing and insuring leases. The sublessee is willing to surrender his present lease provided that he receives appropriate compensation. The freeholder has also agreed to the redevelopment, provided that the ground rent is adjusted on completion of the development to $7\frac{1}{2}$ per cent of the rack rental value of the new development, and is reviewed every 5 years.

The following information is also available

  (i) The rateable value of the existing premises is £1 200.
 (ii) Estimated building cost £500 per m^2.
(iii) Period of building 1 year.
(iv) Similar offices, totalling 700 m^2 net, have been let recently on a 5 year full repairing and insuring lease at a rent of £46 000 per annum, subject to the payment of £20 000 on entry.

Last week, the local planning authority's revocation of the planning permission was confirmed.

Prepare a compensation claim for your client.
7.2. The level of compensation for refusal of planning permission for new development is subject to values established over 30 years ago.

Explain, and construct examples to illustrate your answer.
7.3. Explain the basis of compensation in relation to;

(a) The discontinuance of the existing use of a building and/or land in accordance with *Sections 170 and 178 of the Town and Country Planning Act 1971*. Construct an example to illustrate your answer.

(b) The refusal of planning permission for the enlargement of a block of flats.
7.4. Your client is the freehold owner−occupier of waste land, which has an existing use value of £10 000. He is, also, lessee of an adjacent site comprising house

and garden, the lease having 5 years unexpired at a ground rent of £50 per annum. The house has a rateable value of £220 and a vacant possession value of £40 000. Your client is negotiating to buy the house under the terms of the *Leasehold Reform Act, 1967* and subsequent legislation, but the enfranchise-ment price has not yet been agreed. The two sites cannot be developed separately, so your client intends to demolish the house and combine the sites. He holds planning permission to build 2500 m² net of industrial accommodation but the local planning authority's revocation order has just been confirmed.

Prepare a compensation claim on behalf of your client.

The following information is available:

 (i) Building period 1 year.
(ii) Estimated building cost £250 m².
(iii) Newly built industrial accommodation, having a net area of 3000 m² was recently let on a 5 year full repairing and insuring lease at a rent of £107 000 per annum, subject to the payment of a premium of £40 000 on entry.

7.5. Some years ago, a freehold developer of land obtained outline planning consent for 100 bungalows. He later obtained full planning consent, but the local plan-ning authority has recently revoked the planning permission. The district council has offered to buy the site for £50 000 otherwise it will remain un-developed.

You are required to prepare a compensation claim on behalf of the developer.

The following information has been established had the development been carried out:

  (i) Building period 18 months.
 (ii) Sale price £80 000. 25 bungalows could be sold after 9 months, 50 after 12 months and the remaining 25 after 18 months.
(iii) Gross floor area of each bungalow 80 m².
(iv) Building costs (inclusive of siteworks and contingencies) 550 per m².
 (v) Retention fund 5 per cent — $2\frac{1}{2}$ per cent to be released on completion and $2\frac{1}{2}$ per cent 6 months after completion.
(vi) Quantity surveyor's and architect's fees — 10 per cent of building costs (50 per cent to be paid on commencement of building, 25 per cent on completion and 25 per cent 6 months after completion).
(vii) Estate agency, legal fees and advertising 4 per cent of sale price.
(viii) Financing 3 per cent per quarter.

It is required that as part of your compensation claim you produce a 3 monthly discounted cash flow.

7.6. The freehold owner-occupier of a shop in the main street of a provincial town, frontage 7 m and depth 20 m, has applied for planning permission to extend it at the rear by 2 metres. His planning application has been refused.

A similar shop, frontage 8 m and depth 17 m, has recently let on full repair-ing and insuring terms at a rent of £14 800 per annum.

Prepare a compensation claim on behalf of the freeholder.

7.7. Explain the basis of compensation in each of the following situations:

(a) A detached house is a listed building because of 'its special architectural and historic interest'. The freehold owner–occupier wishes to extend the house at the rear, but has been refused listed building consent.

(b) A freehold developer has applied for planning permission to develop woodland for residential use. This has been refused because the woodland is subject to a tree preservation order.

PLANNING COMPENSATION – SUGGESTED ANSWERS

Question 7.1

It is assumed that the revocation order has become effective before formal negotiations to remove the existing tenant have commenced and development has not begun. Any claim would be governed by *Section 164 of the Town and Country Planning Act 1971*.

The first step is to obtain a rent per m^2 for offices by analysing the similar transaction. The premium should be analysed from both freehold and leasehold viewpoints

i.e. Rent passing £ 46 000 pa

Plus Annual Equivalent of Premium:

Freeholder

$$\frac{£20\,000}{\text{YP 5 years at 8 per cent}} = \frac{£20\,000}{3.993} = £5\,008$$

Lessee

$$\frac{£20\,000}{\text{YP 5 years at 9 per cent and 3 per cent}} = \frac{£20\,000}{2.476} = £8\,077$$
$$\text{(tax 40 per cent)}$$

Average say £ 6 540 pa

£52 540 pa

say £52 500 pa

$$\text{Rent per m}^2 = £\,\frac{52\,500}{700} = £75 \text{ per m}^2$$

It is now necessary to prepare a claim for the loss in land value due to the revocation order. This is the difference in values between the value with the existence of planning permission and with the effect of revocation.

Value with effect of planning permission

Net rental value = 1200 m^2 × £75		£ 90 000
YP 50 years at 7 per cent [see note 1]		13.8
Gross Development Value		£1 242 000

Less

(i) Compensation to sublessee
[see note 2]
(a) Value of lease

Net rental value	£3 000 pa
Rent paid	£2 000 pa
Profit rent	£1 000 pa

YP 3 years at 9 per cent
and 3 per cent
(tax 40 per cent)   <u>1.589</u>
£ 1 589

(b) 6 × Rateable value (£1 200) £ 7 200

(c) Surveyor's and legal fees
say 3 per cent   £ 263

(d) Financing 1 year at
12 per cent   <u>£ 1 086</u>

£10 138
say   £ 10 200

(ii) Demolition [see note 3]   say   £ 5 000
(iii) Building costs – 1 200 m^2 + 10 per cent
[see note 4] = 1 320 m^2 × £500   = £660 000
(iv) Contingencies 10 per cent on items
(ii) and (iii) [see note 5]   = £ 66 500
(v) Quantity surveyor's and architect's
fees – 10 per cent on items (ii),
(iii) and (iv) [see note 6]   = £ 73 150
(vi) Financing 1 year at 12 per cent on
half items (ii) to (v) [see note 7]   = £ 48 280
(vii) One year's rent [see note 8]   = £ 200
(viii) Revised ground rent pa
[see note 9]   £ 6 750 pa
YP 49 years at 7 per cent   13.767
× PV of £1 in 1 year at
7 per cent   <u>0.935</u>

£86 886
say   £ 86 900

(ix) Estate agency, legal fees say 3 per cent
      of Gross Development Value say   = £ 37 300
 (x) Developer's Profit [see note 10] say
      10 per cent of Gross Development Value   = £124 200
                                               _____
                                               £1 111 730
                                               _____
Value of land, acquisition costs and finance =   £   130 270
[see note 11]

If acquisition costs are 4 per cent, and financing is 12 per cent, then the value of
the land with effect of planning permission is

$$\frac{£130\,270}{1.1648} \text{ [see note 11]} = \text{say £111 800}$$

Notes
1: Because of the long term (50 years), the rent has been capitalised by a single
   rate Years' Purchase. An alternative would be to use a dual rate tax-adjusted
   Years' Purchase.
2: The sublessee has been compensated with the value of the unexpired term of
   the lease and the statutory entitlement of 6 x Rateable value. In practice, the
   sublessee would probably realise he could ask for a much greater amount to
   expedite his departure!
3: The demolition of the factory has been assumed to cost £5 000.
4: The net floor area of 1200 m^2 has been increased by 10 per cent to produce a
   gross floor area of 1320 m^2.
5: A contingency allowance of 10 per cent has been allowed on demolition and
   building costs.
6: Professional fees have been calculated at 10 per cent on demolition, building
   costs and contingencies (the last item may be debatable, but it may be argued
   that if the contingency allowance is used, fees would be paid upon it).
7: This item illustrates a shortcoming of the residual method of valuation. It is
   assumed that costs are evenly distributed throughout the building period, so
   that financing is allowed on 50 per cent of the costs. In reality, costs are
   variable on a monthly basis throughout the building period.
8: This is the ground rent payable during the building period.
9: The revised ground rent of £6 750 per annum is $7\frac{1}{2}$ per cent of the rental value
   of £90 000.
10: As part of a typical residual valuation, developer's profit has been included at
    10 per cent of Gross Development Value. This is pre-tax.
11: Although the head lessee is in occupation, acquisition costs and financing are
    deducted from the residue, as a prospective purchaser would incur these
    charges. 1.1648 is the value of the land and costs (1.04) with interest at 12 per
    cent on it, i.e. 1.04 x 1.12 = 1.1648.

Value with the effect of the revocation order

Existing Use Value [see note 1]

Rent received	£2 000 pa	
Less Ground rent	£ 200 pa	
Profit rent	£1 800 pa	
YP 3 years at 9 per cent	2.531	£ 4 556
Current rental value	£3 000 pa	
Less Ground rent	£ 200 pa	
Profit rent	£2 800 pa	
YP 47 years at 9 per cent	10.918	
× PV of £1 3 years at 9 per cent	0.772	£23 600
		£28 156
		say £28 200

Note

1: It is assumed that the sub-lessee would occupy until the expiration of his lease.

$$\text{Claim} = £111\,800 - £28\,200 = £83\,600$$

In addition, there would be a claim for 'abortive expenditure' which in this situation might include professional fees incurred at this stage, and damages to the contractor if a building contract has been established.

The authority would resist a claim for surveyor's and architect's fees incurred for obtaining outline planning permission and loss of profits.

Question 7.2

This answer should explain how a claim was established in relation to land having development potential under *Part VI of the Town and Country Planning Act 1947* and how this subsequently extended to become *"an unexpended balance of established development value."* (UXB)

i.e. Under the 1947 Act, it became necessary for owners of development land to obtain planning permission for new development.

If planning permission was granted, a development charge of 100 per cent of the gain had to be paid. If the planning permission was refused, those owning land could not realise its development potential. With this in mind, the Central Land Board was established with a global fund of £300m. Freeholders and leaseholders of development land could submit claims equivalent to the latent development value as at 1 July 1948. This was calculated as being the difference between unrestricted development value and restricted value.

Example

Land has existing use value of £1 000 and development value of £8 000, so that the Part VI claim would have been £7 000.

If the owner applied for planning permission and it was granted the Part VI claim was extinguished; if it was refused, the £7 000 was paid.

If the owner sold the land, it was expected that he would sell at existing use value, would receive £7 000 from the Central Land Board, and the new purchaser/developer would pay a development charge of £7 000.

This might be illustrated diagrammatically:

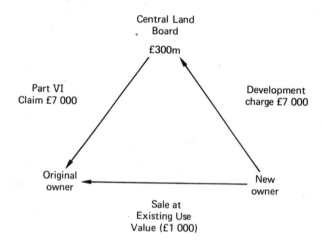

This arrangement meant that the control of development value was firmly in the hands of the Central Land Board.

In 1951 the development charge was abolished and payments would only be made on claims established between 1948 and 1951. After *The Town and Country Planning Act, 1954*, the claim had $\frac{1}{7}$th added to it to represent interest from 1948 to 1954, i.e. £7 000 + $\frac{1}{7}$th = £8 000.

This figure is termed the *'unexpended balance of established development value'* and provides the maximum amount of payment thereafter. *Sections 140–143 of the Town and Country Planning Act 1971* prescribe ways in which the UXB may be reduced or extinguished:

(i) where there is a planning refusal and compensation is paid
(ii) where new development is permitted on the land
(iii) where a payment is made in a compulsory purchase situation based on development value of the land or for severance or injurious affection.

The only compensation now payable in relation to a refusal of a planning permission on land would be limited to the existence of a UXB, i.e. no UXB — no compensation. (The only remedy available to an aggrieved applicant would be to serve a purchase notice if he could prove that his land was 'incapable of reasonable beneficial use'.)

Question 7.3

(a) A local planning authority may require the use of land and/or buildings to be discontinued "if it appears . . . that it is expedient in the proper planning of (their) area" (*Section 51, Town and Country Planning Act 1971*) *Section 170* provides for compensation and this may be under three heads of claim:

(i) Loss in land value. This would be the difference between existing use value and the value with the effect of discontinuance, taking into account any alternative planning permission.
(ii) Disturbance. This could involve the extinguishment of a business with redundancy payments.
(iii) Expenses incurred.

Example

A discontinuance order has been served on the freehold owner–occupier of a chemical factory, which is located in a predominantly residential area. The factory has 500 m² of lettable floor space.

The site could have planning permission for two residential building plots, each worth £15 000.

Typical claim

(i) Loss in land value:		
500 m² at say £15 per m²		= £ 7 500 pa
YP in perpetuity at 10 per cent		10
Existing Use Value		£ 75 000
Less Value of building plots	£30 000	
Less Demolition	say £ 5 000	£ 25 000
Loss in land value		£ 50 000
(ii) Disturbance:		
Extinguishment of business		
[see note 1]		
Average net profits	say £20 000 pa	
YP	say 2	
		£ 40 000
Redundancy payments [see note 2]		say £ 10 000
(iii) Incidental expenses [see note 3]		say £ 5 000
	Total claim	£105 000

Notes

1: The disturbance claim has been prepared in the same format as a compulsory purchase claim under *Rule 6 of the Land Compensation Act 1961*. The average net profits are based on a minimum of 3 years' figures. Care must be taken by the claimant to make a case for ignoring those years where figures may have been reduced because of the threat of discontinuance.

A YP of 2 would reflect a steady increase in real profitability, trying to disregard inflation.

2: Redundancy payments would be determined by statutory entitlements related to number of employees, sex, age and years of service.

3: This might include loss on enforced sale of fixtures, fittings and stock.

(b) It is initially important to establish whether the enlargement falls within *Part II of the Eighth Schedule, Town and Country Planning Act 1971*. This allows for tolerances in accordance with subsequent Development Orders, currently an increase of 10 per cent. In these cases, refusal of planning permission may lead to compensation claims being the difference in value between the value with Schedule 8 rights and existing use value. However, the position in relation to enlargement of flats is different. In 1983, the Court of Appeal reversed a Lands Tribunal decision in *Camden LBC v. Peaktop Properties (Hampstead) Ltd.* that compensation was payable under Section 169 where the local planning authority had declined to grant planning permission for the extension of flats. The case involved two buildings each of 104 flats, and the developers applied to provide an additional floor giving an extra 11 flats per building. The volume did not increase by more than 10 per cent but the floor space was increased by 11.4 per cent.

The existence of these provisions would be an encouragement to developers to submit applications to build upwards, knowing that if planning permission was refused full compensation would be paid. *The Town and Country Planning (Compensation) Act 1985, Section 2* seeks to strengthen the Court of Appeal decision, namely Schedule 8 is not to apply to the enlargement of a building in existence on (1 July 1948) if:

(a) the building contains two or more separate dwellings divided horizontally from each other or from some other part of the building, and
(b) the increase would result in either an increase in the number of dwellings contained in the building or increase of more than 10 per cent cubic content in any such dwelling contained in the building.

This appears to have eliminated the liability of compensation payments for the enlargement of blocks of flats.

Question 7.4

The first step is to obtain a rent per m² from the analysis of similar industrial accommodation. The premium should be analysed from both freehold and leasehold viewpoints

i.e. Rent passing   £107 000 pa
 Plus Annual equivalent of Premium

Freeholder

$$\frac{£40\,000}{\text{YP 5 years at 8 per cent}} = \frac{£40\,000}{3.993} = £10\,018$$

Leaseholder

$$\frac{£40\,000}{\substack{\text{YP 5 years at 9 per cent and 3 per cent} \\ \text{(tax 40 per cent)}}} = \frac{£40\,000}{2.476} = £16\,155$$

                                            Average say   £ 13 086 pa
                                                                 £120 086 pa
                                                  say   £120 000 pa

$$\text{Rent per m}^2 = \frac{120\,000}{3\,000} = £40 \text{ per m}^2$$

It is necessary to prepare a claim for loss in land value due to the revocation order. This is the difference in values between the value with the existence of planning permission and with the effect of revocation.

Value with effect of planning permission

Net rental value = 2 500 m^2 x £40 =			£100 000 pa
YP in perpetuity at 8 per cent			12.5
Gross Development Value			£1 250 000

Less

(i) Enfranchisement Price [see note 1]

Ground rent	£50 pa		
YP 5 years 6 per cent	4.212	£ 211	

Reversion to Section 15 rent:
Site value say 25 per cent of vacant
possession value = 25 per cent of
£40 000 = £10 000

6 per cent return =	£600		
YP in perpetuity deferred 5 years at 6 per cent	12.454	£ 7 472	
		£ 7 683	
say		£ 7 700	
Acquisition fee say 4 per cent		£ 308	
1 year's financing at 12 per cent		£ 960	
		£ 8 968	
	say	£ 9 000	

(ii)	Demolition [see note 2]	say	£ 2 000	
(iii)	Building costs 2 500 m^2 + 10 per cent [see note 3] = 2 750 m^2 gross x £250	=	£ 687 500	
(iv)	Contingencies [see note 4] 10 per cent on items (ii) and (iii)	=	£ 68 950	
(v)	Quantity Surveyor's and Architect's fees [see note 5] 10 per cent on items (ii)-(iv)	=	£ 75 845	
(vi)	Financing 1 year at 12 per cent for half of items (ii)-(v) [see note 6]	=	£ 50 060	
(vii)	Estate agency, legal fees say 3 per cent of Gross Development Value	=	£ 37 500	
(viii)	Developer's Profit [see note 7] 10 per cent of Gross Development Value	=	£125 000	£1 055 855
	Value of land acquisition costs and financing [see note 8]	=		£ 194 145

If acquisition costs are 4 per cent and financing is 12 per cent, then the value of the land with effect of planning permission is

$$\frac{£194\,145}{1.1648} \text{ [see note 8]} = \text{say } £166\,700.$$

Notes

1: The enfranchisement price is calculated for a house having a rateable value less than £500 in the provinces and follows the guidelines established in *Farr v. Millersons Investments Limited (1971)*. The Section 15 ('modern ground') rent has been calculated using the standing house approach with the assumption that site value is 25 per cent of entirety value. This has then been decapitalised at 6 per cent to obtain the rent.

2: Demolition costs are assumed to be £2 000. This might depend on the value of recoverable material.

3: The net floor area of 2500 m² has been increased by 10 per cent to produce a gross floor area of 2750 m².

4: Contingencies have been provided at 10 per cent of demolition and building costs.

5: Quantity Surveyor's and Architect's fees have been allowed at 10 per cent of demolition and building costs and contingencies.

6: As in Question 7.1, financing has been provided on the assumption that costs distribute evenly during the building period.

7: Developer's profit has been included at 10 per cent of Gross Development Value (pre-tax).

8: See answer 7.1, note 11.

Value with the effect of the revocation order

Existing Use Value [see note 1]

Value of waste land	£10 000
Value of house	£40 000
	£50 000
Less Enfranchisement price, acquisition fee and financing	£ 9 000
	£41 000

Note

1: It is assumed that the client would carry out enfrancisement and become freeholder of the house worth £40 000. He would still own the waste land worth £10 000.

$$\text{Claim} = £166\,700 - £41\,000 = £125\,700$$

There would also be a claim for 'abortive expenditure', i.e. those items of expenditure rendered useless by the decision, as explained in answer 7.1.

Question 7.5

As with earlier questions, the claim entails the calculation of the land value with the effect of planning permission and the deduction from this of the land value with the effect of revocation. The latter figure will be equivalent to the district council offer of £50 000.

It is necessary to calculate the items, which would be incorporated in a residual method and then distribute them into a quarterly pa discounted cash flow. i.e.

(i) Value of 25 bungalows = 25 × £80 000 = £2m

(ii) Value of 50 bungalows = 50 × £80 000 = £4m

(iii) Estate agency and legal fees on 25 bungalows at 4 per cent = £80 000 and on 50 bungalows = £160 000

(iv) Building cost of 100 bungalows = 100 × 80 m² × £550 per m² = £4.4m
If this is distributed evenly over 6 quarters = £733 333 per quarter.

(v) The provision of retention fund is a system whereby 5 per cent of building costs (and, presumably, contingencies if used) is retained each time payments are made. This is then released to the building contractor − $2\frac{1}{2}$ per cent on completion of the works and $2\frac{1}{2}$ per cent after 6 months.

Total retention = 5 per cent of £4.4m = £220 000, i.e. £36 666 per quarter.

(vi) Quantity Surveyor's and Architect's fees 10 per cent of building costs and contingencies = £440 000, distributed as indicated in the question.

(vii) Developer's Profit will be allowed at 10 per cent of the value of the bungalows, i.e. £800 000 and incorporated in the cash flow at 18 months.

Value with the effect of planning permission

Item	Immediate	End of months						
		3	6	9	12	15	18	24
Income				+2m	+4m		+2m	
Estate agency and legal fees				– 80 000	– 160 000		– 80 000	
Building costs and contingencies		–733 333	–733 333	– 733 333	– 733 333	–733 333	–733 333	
Retention		+ 36 666	+ 36 666	+ 36 666	+ 36 666	+ 36 666	+ 36 666	
QS's and Architect's fees	–220 000						–110 000	–110 000
Release of retention							–110 000	–110 000
Developer's profit							– 800 000	
Total	–220 000	–696 667	–696 667	+1 223 333	+3 143 333	–696 667	+203 333	–220 000
Discounted at 3 per cent per quarter [see note 1]	1.0	0.97	0.943	0.915	0.888	0.863	0.837	0.789
Total flow	–220 000	–675 767	–656 957	+1 119 350	+2 791 280	–601 224	+170 190	–173 580

Land value and acquisition costs =

$$\text{Inflows} - \text{Outflows} = 4\,080\,820 - 2\,327\,528 = \text{£1}\,753\,292$$

$$\text{Land value} = \frac{\text{£1}\,753\,292}{1.04} = \text{say £1}\,685\,858$$

Note

1: PV of £1 at 3 per cent per quarter is obtainable from *Parry's Valuation Tables*. The figures used as '*n*' in the formula

$$\frac{1}{(1+i)^n}$$

may be regarded as interest periods rather than years or months.

So that claim	=	£1 685 858
Less	£	50 000
		£1 635 858
say		£1 635 900

As before, there would be an additional claim for abortive expenditure.

Question 7.6

The proposed extension of 14 m^2 does not exceed 10 per cent of the existing area of 140 m^2 (nor presumably 10 per cent of the cubic capacity) and is, therefore, 'permitted development' under *Schedule 8, Part II of the Town and Country Planning Act 1971*. The compensation entitlement will be the difference between the value with planning permission and the value as existing. The first step is to analyse the rental transaction of the similar shop to obtain a Zone A rent per m^2.

Assume 5 metre zones, and 'halving back' with two 5 metre zones and a remainder.

Let rental value per m^2 Zone A = £X

Thus

Zone A	= 8 m x 5 m x	X	= 40X
Zone B	= 8 m x 5 m x $\frac{1}{2}X$		= 20X
Remainder	= 8 m x 7 m x $\frac{1}{4}X$		= 14X
	Rental value		= 74X

But Rental value = £14 800
. and X = £ 200 per m^2

The information is now applied to the subject property. Zone depths used should be the same as those adopted to analyse the comparable transaction.

Zone A   = 7 m × 5 m × £200 = £ 7 000
Zone B   = 7 m × 5 m × £100 = £ 3 500
Remainder = 7 m × 10 m × £ 50 = £ 3 500

                Rental value   = £14 000

Value with the effect of planning permission

Existing rental value per annum	= £ 14 000 pa
Extension — 7 m × 2 m × £50 [see note 1]	£ 700
Full rental value	£ 14 700 pa
YP in perpetuity at 7 per cent	14.28
	£209 916
Less Cost of extension —	
7 m × 2 m × £300 per m^2 [see note 2]	= £ 4 200
	£205 716
Value with planning permission say	£205 700

Notes
1: It is assumed that the extension has a rental value per m^2 of £50, the value of the rear zone of the original building.
2: Building costs of £300 per m^2 have been assumed, to include cost, contingencies, fees and financing.

Value as existing

Existing rental value as calculated	= £ 14 000 pa
YP in perpetuity at 7 per cent	14.28
	£199 920
Value as existing say	£199 900

So that compensation claim =
£205 700 — £199 900   = £5 800

Question 7.7

(a) The Secretary of State is empowered to compile a list of buildings of special architectural and historic interest and in order to extend such a building, as in this case, listed building consent would be required in writing from the local planning authority or the Secretary of State.

If listed building consent is refused, *Section 171, Town and Country Planning Act 1971* may apply if the refusal relates to 'permitted development' or the proposed work was not development at all.

In the case of the extension to the house, this may be regarded as 'permitted development' so that the compensation claim will be the difference between the value of the house with the extension and existing use value.

e.g *Value with consent*

Value of house and extension	£100 000
Less Cost of extension	£ 10 000
	£ 90 000
Less Value as existing	£ 75 000
Claim	£ 15 000

Note

1: Purchase notice provisions in relation to listed buildings are the same as in general cases — the owner must be able to show that the property is 'incapable of reasonable beneficial use'.

(b) *Section 174, Town and Country Planning Act 1971* deals with compensation in relation to refusals of consent where trees are subject to tree preservation orders. The developer should apply for consent to fell the trees and if this is refused, he has a compensation claim.

He must value the land with the development value which could have been expected had the tree preservation order not existed. The existing use value will then be deducted from this to establish the compensation claim.

e.g. Value of development land	say	£500 000
Less Value of woodland	say	£ 50 000
	Claim	£450 000

It may be necessary to calculate the development value by the use of a residual method or discounted cash flow techniques as shown in earlier examples.

8 Asset Valuation

Valuations of fixed assets to be shown in financial statements are required for purposes such as the annual accounts of companies as specified in the *Companies Act 1967*, takeover bids, flotations, company borrowing, capital re-organisation, insurance, taxation, liquidation, bankruptcy and receiverships.

The valuer is often required to produce values for land and depreciable assets such as buildings and plant and machinery, and a knowledge of current cost accounting principles is essential. This requires an assessment of current cost or current value of assets including real property rather than historic or acquisition costs, which might have prevailed at the time the property was built.

The Royal Institution of Chartered Surveyors has an Assets Valuation Standards Committee which has produced *Guidance Notes*, and the Incorporated Society of Valuers and Auctioneers produced a *Guide to Asset Valuations* in 1988.

There are also *Statements of Standard Accounting Practice* (SSAPs) issued by a body known as the Accounting Standard Committee (ASC), and these are mandatory on members of the accounting bodies that form the ASC.

The Royal Institution of Chartered Surveyors published new *Guidance Notes* in 1989, and an important section is entitled 'Additional Valuations made on Special Assumptions' (GN28). In exceptional circumstances it may be necessary for a valuer, in order to advise fully upon the potential value of a property, to report an additional valuation based on a 'special assumption'. This is defined as an assumption which — in the actual circumstances prevailing in the market at the time of the valuation — could not reasonably be expected by the valuer to be made by a prospective purchaser who is not a purchaser with a special interest. A valuer may provide an additional valuation on a property based on the assumption that the purchaser will, in the future, be able to obtain planning permission for redevelopment. The valuation of a portfolio as a whole or in parts may produce a greater or lesser figure than the aggregate value of the separate properties it contains, and this may be recognised. Special assumptions can be made only if the property is likely to be the object of a special purchaser. However, the price which a non-special purchaser would pay might include some part of that marriage value (to a special purchaser) because of the hope that the non-special purchaser might eventually be able to resell to the special purchaser at an inflated price. Hope value and marriage value may, therefore, be included in open market value.

Usually, the valuation of land and buildings will be on the basis of Open Market Value, as defined in RICS Guidance Notes:

"the best price at which an interest in a property might reasonably be expected to be sold by private treaty at the date of valuation assuming

(a) a willing seller;

(b) a reasonable period within which to negotiate the sale taking into account the nature of the property and the state of the market;

(c) the values will remain static throughout the period;

(d) the property will be freely exposed to the market;

(e) no account is to be taken of an additional bid by a special purchaser."

Examination questions will often require the calculation of Open Market Value (OMV), and the dividing of this figure between land value ('residual amount') and buildings and, possibly, plant and machinery ('depreciable amount'). It may also be necessary to calculate Alternative Use Value. The valuer must also establish into which of the following categories his valuation fits:

(i) Land and buildings in owner–occupation (non-specialised properties)

(ii) As (i) (specialised properties)

(iii) Land and buildings surplus to the business

(iv) Land and buildings held as investments or for development

(v) Property in the course of development

(vi) Leaseholds.

Questions may also deal with depreciation, defined in Statement of Standard Accounting Practice (SSAP) 12 as "the measure of the wearing out, consumption or other loss of value of a fixed asset whether arising from use, effluxion of time or obsolescence through technological and market changes."

Where depreciation is to be taken into account, an estimation of the useful life of the asset is required and the method of 'writing down' the asset. Straight-line depreciation seems to be commonly used, although sum-of-the-years digits and declining balance of fixed percentage methods are, perhaps, more realistic. After all, the greatest depreciation happens in buildings towards the end rather than at the beginning of useful life.

However, a large concern, British Gas plc., use depreciable periods of up to 50 years for freehold and leasehold buildings and 5 to 20 years for plant and machinery, and depreciate on a straight-line basis.

ASSET VALUATION – QUESTIONS

8.1. Value the following for inclusion in company accounts, calculating the residual and depreciable amounts:

A single-storey freehold factory in owner–occupation, having 1000 m² of net floor space, built 15 years ago with an anticipated economic life of 50 years from new.

You have obtained information that similar premises, 1500 m² net, have recently let on a 5 year full repairing and insuring lease at a rent of £32 350 per annum subject to the payment of a premium of £15 000 on entry.

The site could be redeveloped as follows:

A factory, 2000 m² net floor space, could be built within 1 year, giving an estimated rent per annum of £40 per m² on full repairing and insuring terms. Costs including building, contingencies, financing and fees are considered to be £250 per m².

8.2. Value the following for inclusion in company accounts, calculating the residual and depreciable amounts:

A three-storey freehold warehouse in the Midlands in owner–occupation, having 900 m² of net floor space, built 30 years ago with an anticipated economic life of 40 years from new.

You have obtained the following information:

(i) The lessee of a similar warehouse, having 1000 m² of net floor space, has recently surrendered the 2 years' remaining of a full repairing and insuring lease at a rent of £10 000 per annum, and has taken a 10 year full repairing and insuring lease at a rent of £13 400 per annum with a 5 year rent review.

(ii) The replacement cost of a similar warehouse including building, contingencies, fees and finance would be £275 per m².

8.3. The statement of Standard Accounting Practice (SSAP) 12 recommends that "depreciation should be allocated to accounting periods so as to charge a fair proportion to each accounting period during the expected useful life of the asset."

Describe the possible methods of allowing for depreciation on property assets, using examples to illustrate.

8.4. A large printing company requires a valuation of a four-storey workshop and offices situated in a local city centre for inclusion in its corporate accounts.

The building is pre-war and comprises 3000 m² net of printing workshops on the upper floors and 1000 m² net of ground floor offices. Tenure is leasehold with 40 years unexpired at a fixed ground rent of £500 per annum. Planning permission for use of the entire building as offices has been obtained. It is estimated that conversion and some refurbishment would cost £300 000 if the company was to relocate its printing operations. The refurbished offices would let at £50 net per m².

Making reasonable valuation assumptions, prepare valuations for inclusion in a report to the company.

ASSET VALUATION – SUGGESTED ANSWERS

Question 8.1

The first step in answering this question is to analyse the recent letting of similar premises to obtain a rent per m². The premium should be analysed from both free-hold and leasehold viewpoints.

i.e. Rent passing   £32 350 pa
Plus Annual Equivalent of Premium:

$$\text{Freeholder £} \ \frac{15\,000}{\text{YP 5 years at 10 per cent}} = £\ \frac{15\,000}{3.79} = £3\,958$$

$$\text{Lessee £} \ \frac{15\,000}{\substack{\text{YP 5 years at 11 per cent and 3 per cent} \\ \text{(tax 40 per cent)}}} = £\ \frac{15\,000}{2.359} = £6\,358$$

	Average	£ 5 158 pa
		£37 508
	say	£37 500 pa

$$\text{Rent per m}^2 = \frac{37\,500}{1\,500} = \qquad £\quad 25$$

Applying to the subject property:

1000 m² x £25	=	£25 000
YP in perpetuity at 10 per cent	=	10
OMV [see note 1]		£250 000

Note

1: Some valuers add acquisition costs, stamp duty and valuers fees to this figure, but it may be better practice to show these separately in the accounts.

Having calculated the Open Market Value, it is then necessary to divide this between the residual and depreciable amounts. Because the value of the land is not available, the depreciated replacement cost must be used as follows:

Gross replacement cost =
1 000 m^2 + 10 per cent [see note 1]
= 1 100 m^2 gross × £250 [see note 2] =   £275 000

Depreciated replacement cost

[see note 3] $= \dfrac{\text{Remaining years}}{\text{Total years}} =$

$$\frac{35}{50} \times £275\,000 \quad = \qquad £192\,500$$

So that the Residual amount is

$$OMV - Drc = £250\,000 - £192\,500$$
$$= £\ \ 57\,500$$

and the Depreciable amount is £192 500 [see note 4]

Notes
1: The net floor area has been grossed up by 10 per cent to obtain a gross floor area. This percentage may vary according to the property type.
2: The £250 per m^2 is a composite figure; in many instances, a residual method of valuation would be required.
3: A straight-line method of depreciation has been used.
4: These figures would be included in the appropriate company accounts after agreement between the valuer and the directors.

Alternative use must be considered, even though the existing land and buildings have been valued as a going concern. The alternative use value should be calculated and included in the Directors' Report which accompanies the accounts.

2 000 m^2 × £40		= £ 80 000 pa
YP in perpetuity at 10 per cent		10
Gross Development Value		£800 000
Less		
Demolition say	say £ 10 000	
Costs 2 200 m^2 × £250 [see note 1]	= £550 000	
Estate agency and legal fees say		
3 per cent of GDV	= £ 24 000	
Developers Profit say 10 per cent of GDV	= £ 80 000	£664 000
		£136 000

Assuming acquisition fees at 4 per cent and financing at 12 per cent per annum, then let land value = X.

$$£136\,000 = 1.04X + (1.04X)\,0.12$$
$$£136\,000 = 1.04X + 0.1248X$$
$$£136\,000 = 1.1648X$$

$$\text{Land value} = \frac{136\,000}{1.1648} = £116\,750$$

Notes

1: A 10 per cent allowance has been made to obtain a gross floor area. Again, a more detailed residual valuation might be used.
2: The alternative use value of £116 750 and the value of the completed development (£800 000) would be included in the Directors' Report.

Question 8.2

Firstly, it is necessary to carry out a surrender and renewal valuation from both landlord and tenant viewpoints to calculate the current rental value per annum of the similar warehouse, and then ascertain a rent per m² for the subject property.

Freeholder's present interest

Let rental value on full repairing and insuring terms = $£X$

Net income	£10 000 pa	
YP 2 years at 11 per cent	1.713	£17 130
Reversion to full net rental value	$£X$ pa	
YP in perpetuity deferred 2 years at 11 per cent	7.378	£ 7.378X
	Capital value	£17 130 + 7.378 X

Freeholder's proposed interest

Net income	£13 400 pa	
YP 5 years at 11 per cent [see note 1]	3.696	£49 526
Reversion to full net rental value	$£X$ pa	
YP in perpetuity deferred 5 years at 11 per cent	5.395	5.395X
	Capital value	£49 526 + 5.395X

$$\text{Present} = \text{Proposed [see note 2]}$$
$$£17\,130 + 7.378X = £49\,526 + 5.395X$$

$$X = \frac{£49\,526 - 17\,130}{7.378 - 5.395}$$

$$= £16\,340 \text{ pa}$$

Lessee's present interest

Full net rental value	£X	pa
Less Rent paid	£10 000	pa
Profit rent	£X − 10 000 pa	
YP 2 years at 12 per cent and 3 per cent (tax 40 per cent)	1.063	
Capital value	£1.063X − 10 630	

Lessee's proposed interest

Full net rental value	£X	pa
Less Rent paid	£13 400	pa
Profit rent	£X − 13 400 pa	
YP 5 years at 12 per cent and 3 per cent (tax 40 per cent)	2.304	
Capital value	£2.304X − 30 870	

$$\text{Present} = \text{Proposed [see note 2]}$$
$$£1.063X - 10\,630 = £2.304X - 30\,870$$

$$X = \frac{£30\,870 - 10\,630}{2.304 - 1.063}$$

$$= £16\,300 \text{ pa}$$

Averaging the freeholder and lessee's figures, the current net rental value is considered to be say £16 320 pa, i.e. a rent of £16.4 per m^2 (say).

Notes

1: Although the new lease is for a 10 year period, the rent would be the current rental value pa at the 5 year review period.

2: In this type of calculation, it is considered that the landlord should be no worse off and the tenant no better off when the new lease is created.

Applying to the subject property

900 m² × £16.4 =	£ 14 760 pa
YP in perpetuity at 11 per cent	9.09
OMV	£134 170

Again, it is necessary to divide this figure between residual and depreciable amounts.

Gross replacement cost =

900 m² + 10 per cent	=	
990 m² × £275	=	£272 250
Depreciated replacement cost	=	

$$\frac{\text{Remaining years}}{\text{Total years}} = \frac{10}{40} \times £272\,250$$

$$= £68\,000$$

So that Residual amount is

$$\text{OMV} - \text{Drc} = £134\,170 - £68\,000$$
$$= £66\,170$$

Question 8.3

This answer should describe the following:

(a) Straight-line method
(b) Sum-of-the-years' digits method
(c) Declining balance of fixed percentage method.

The methods may be illustrated by constructing a simple example.

For example a factory building, built 30 years ago and having a useful life of 50 years from new, would cost £200 000 to replace.

Using

(a) Straight-line, the depreciable amount would be

$$\frac{\text{Remaining years}}{\text{Total years}} \times £200\,000$$

$$\frac{20}{50} \times £200\,000 = £80\,000$$

The disadvantage of this method is that it assumes uniform 'writing-down' throughout the life of the building, whereas the greater rate of depreciation occurs towards the end of the building's life.

(b) The sum-of-the-years' digits. This attempts to recognise the uneven nature of depreciation by calculating the number of digits contained in the property's previous life with those in its total life since new. A formula will calculate this, namely

$$\frac{a+l}{2} \times n$$

where a = first year number, l = last year number and n = number of years over which the digits accumulated.

$$\text{Previous life} = \frac{1+30}{2} \times 30 = 465$$

$$\text{Total life} = \frac{1+50}{2} \times 50 = 1\,275$$

So that $\left(\dfrac{465}{1\,275}\right)$ th has been used

i.e. 36.5 per cent has been used so that depreciable amount

$$= 63.5 \text{ per cent} \times £200\,000$$
$$= £127\,000$$

(c) Declining balance of fixed percentage. In this method, a fixed percentage is applied to the diminishing balance to calculate the declining amounts of accrued depreciation.

For example, if 4 per cent per annum depreciation was allowed over 30 years, then the depreciable amount would be

$$£200\,000 \times (1-i)^n$$
$$= £200\,000 \times 0.96^{30}$$
$$£200\,000 \times 0.294 = £58\,800$$

SSAP 12 leaves the method of allowance for depreciation to the valuer and directors of the company. Another method may be to make annual allowances at a fixed rate of interest to accumulate to the replacement cost of the building at the end of its life. This may be by a sinking fund provision at an accumulative rate of interest or an annuity provision at a remunerative rate of interest. The problem with these methods is that they will recoup historic cost not the ultimate replacement cost.

Question 8.4

The primary valuation to be undertaken is Open Market Value for existing use on the basis of an investment valuation.

Ground floor offices	− 1 000 m² at £30 [see note 1]	£ 30 000 pa
Upper floor workshop	− 3 000 m² at £15 [see note 2]	£ 45 000 pa
Net rack rental value		£ 75 000 pa
Less Ground rent		£ 500 pa
Net value		£ 74 500 pa
YP 40 years at 11 per cent and 3 per cent (tax 40 per cent)		7.57
OMV		£563 965
		say £564 000

Notes

1: £30 per m² has been assumed to be a reasonable rent for pre-war offices.
2: £15 per m² has been assumed to be a reasonable rent for pre-war printing work-shop. It is appreciated that a comparable for this type of property might be difficult to obtain.

It is necessary to state the Alternative Use Value in the Valuation Certificate if it differs materially from the Open Market Value for existing use.

Net rack rental value = 4000 m² at £50			£ 200 000 pa
YP 38 years at 9 per cent and 3 per cent (tax 40 per cent)			8.764
x PV of £1 in 2 years at 9 per cent [see note 1]			0.842
			£1 475 857
Less (i) Cost of conversion		£300 000	
(ii) Ground rent per annum	£500		
YP 40 years at 9 per cent and 3 per cent (tax 40 per cent) [see note 2]	8.92	£ 4 460	£ 304 460
			£1 171 397
Alternative Use Value say [see note 3]			£1.171m

Notes

1: It is considered that 2 years would be required to carry out the conversion and refurbishment. The yield has been lowered, as the converted property is con-sidered to be a more secure investment than the existing.
2: The ground rent would still be paid for the remaining 40 years.
3: The Alternative Use Value of £1.171m differs materially from the Open Market Value of £564 000, so that it should be included within the Report and Valuation Certificate.

9 Modern Valuation Techniques

Apart from demonstrating an understanding of traditional valuation techniques and the ability to apply these in various circumstances, such as those illustrated in other sections of this book, the student of valuation at this level is required to show some understanding of advances in the theory of valuation. Over the last 25 years or so, there has been increasing awareness that traditional techniques are often less than satisfactory in an inflationary economy. The logic behind these techniques does not always hold true in such an economy and Question **9.2** examines some of the more serious criticisms.

Various alternative valuation techniques have been documented and those considered here are equivalent yield (more an 'amended' traditional approach than an alternative), real value, rational and discounted cash flow. Most valuers are familiar with the latter, which shows explicitly the income flow that is to be valued, including future rental growth. However, calculations can be lengthy and laborious, unless the valuer has computer facilities, and alternative, shorter methods of showing the same calculations have been sought.

The real value method was devised by Dr Ernest Wood, but the method shown here has been described by Dr Neil Crosby in his various articles on the subject as a real value equated yield hybrid. This tends to be somewhat simpler than 'the' real value method and stems from Roy Mason's work following Ernest Wood's approach and Neil Crosby's extension of the concept.

Stephen Sykes and Angus McIntosh conceived the idea of the rational approach and, although there are certain problems attached to this method (explored in Question **9.6**), if applied correctly, it appears to produce satisfactory results.

The majority of formulae used in this section are explained in the text, apart from the derivation of the '3 YPs' used in the real value approach. This is shown as an appendix to this section.

MODERN VALUATION TECHNIQUES — QUESTIONS

9.1. An office property is held from the freeholder on a 42 year lease, with one review after 21 years. The lease was granted 18 years ago, at a rent of £7 500 per annum, on full repairing and insuring terms.

One year ago, the head lessee sublet the property at £20 000 per annum, on full repairing and insuring terms, for the remainder of the term (less 3 days) with reviews every 5 years. The current net rack rental value of the property, on the basis of 5 year reviews, is £22 000 per annum.

Comparable freehold properties, let on the basis of 5 year reviews, are showing initial yields of 5 per cent and gilt edged stock currently yields 10 per cent.

(i) Discuss the major factor to be considered at the rent review between the freeholder and the head lessee in 3 years' time and suggest a possible outcome.
(ii) Discuss the possible difficulties of capitalising the income flow of the head lessee by a traditional approach.
(iii) Value the head leasehold interest using either a discounted cash flow at equated yield or a real value approach, on a gross of tax single rate and a dual rate, tax adjusted basis.

9.2. A prime freehold office property has just been sold for a rack rented capitalisation rate of 6.5 per cent based on 5 year reviews.

A similar property is let on lease with 3 years unexpired, with no further reviews, at £70 000 per annum. Improvements were carried out by the occupying lessee 4 years ago, the provisions of the *Landlord and Tenant Acts* applying.

The current rental value of this property is £95 000 per annum with improvements, but is only £75 000 per annum if they are excluded, both based on 5 year reviews.

Value the freehold and leasehold interests by both real value and conventional techniques, fully annotating your valuations. Discuss the similarities and differences between the two approaches.

9.3. A shop property is currently let on ground lease with 26 years unexpired at a fixed rent of £2 500 per annum. The head lessee sublet the property 7 years ago on a 20 year full repairing and insuring lease with 5 year rent reviews. The current rent passing is £27 000 per annum. The shop is on the ground floor only and measures 6 m frontage by 20 m depth.

Recently, a similar shop nearby was sold for £435 000. Just prior to the sale, the shop was let on a 15 year lease with 3 year rent reviews at £21 750 per annum, on full repairing and insuring terms. This shop measures 5 m frontage by 15 m depth and is also on the ground floor only.

Prepare fully annotated valuations of the freehold and head leasehold interests by an equated yield approach.

9.4. Prepare fully annotated valuations of both the freehold and head leasehold interests in the shop property detailed below, using either a real value or discounted cash flow at equated yield approach.

The property is let at a fixed rent, on full repairing and insuring terms, of £2 000 per annum, on a lease which commenced 19 years ago and was for a term of 42 years. There is provision for one rent review to open market value in 2 years' time.

The property is sublet on a 20 year full repairing and insuring lease, with 5 year reviews, which has 17 years unexpired. The present rent passing is £28 000 per annum and the current full net rental value is estimated to be £37 500 per annum on the basis of 5 year reviews.

A similar freehold property, just let at its full net rental value of £42 000 per annum on 5 yearly reviews, has recently sold for £700 000.

9.5. A freehold shop is let on a full repairing and insuring lease with 14 years unexpired. The lease has 3 yearly rent reviews, the current rent passing is £23 500 per annum and the next review is in 2 years' time. The property measures 6 m frontage by 17.5 m depth and is on the ground floor only.

Value the freehold interest using traditional, equivalent yield and real value approaches, discussing similarities, differences and any problems which may be encountered in the application of the various methods.

You are aware of the following information relating to properties, similar in all respects to the subject property, apart from size:

Property 1   Recently let at its full rental value on full repairing and insuring terms on the basis of 5 year reviews, of £31 500 per annum and subsequently sold freehold for £630 000.
Dimensions are 7 m frontage by 16.5 m depth.

Property 2   Let on full repairing and insuring terms with 4 years unexpired at a fixed rent of £15 750 per annum. The freehold interest has just been sold for £495 000.
Dimensions are 6 m frontage by 16 m depth.

You may assume that the yield on long dated gilts is 10 per cent.

9.6. A shop property is let on ground lease from the freeholder, with 18 years unexpired at a fixed rent of £14 000 per annum.

The property is sublet on a full repairing and insuring lease with 18 years unexpired (less 3 days). The sublease is on 5 year reviews, the next review being in 3 years' time. The current rent payable under the sublease is £45 000 per annum and the current net rack rental value, on the basis of 5 year reviews, is £65 000 per annum.

Freehold interests in similar rack rented properties sell for capitalisation rates of 5 per cent when let on the basis of 5 year reviews.

Long dated gilts yield 10 per cent.

Value the freehold and head leasehold interests by the real value and rational approaches.

Discuss the similarities and differences in approach, of the two techniques, and suggest, giving reasons, improvements to either or both valuation models.

Advanced Valuation

MODERN VALUATION TECHNIQUES – SUGGESTED ANSWERS

Question 9.1

(i) The major factor to be considered at rent review is the level of rent that may be agreed between the freeholder and head leaseholder. There are at present 24 years of the head lease unexpired and, with only one rent review in 3 years' time, it will be desirable, from the landlord's viewpoint, for the rent agreed to compensate for the fact that this rent will be fixed for the following 21 years. What must be decided is the rent for this 21 year period that will also be acceptable to the head leaseholder.

In order to consider the possible outcome, the available information must be analysed in order to convert the current rental value with reviews every 5 years on to the basis of a rent fixed for 21 years.

To achieve this conversion, it is necessary to determine the equated yield, the rate of rental growth per annum and the inflation risk free yield.

The question states that gilts are currently yielding 10 per cent, therefore an equated yield of 12 per cent will be adopted. This follows accepted practice of adding 2 per cent to the yield from gilts to obtain the equated yield for a freehold interest (see Question 9.2). The freehold equated yield is used, since it is the initial yield from comparable freehold properties that is to be analysed in order to determine the implied rate of rental growth per annum.

The implied growth rate is derived using the formula:

$$(1 + g)^n = \frac{\text{YP in perpetuity at } i - \text{YP } n \text{ years at } e}{\text{YP in perpetuity at } i \times \text{PV of £1 in } n \text{ years at } e}$$

where

g = annual rate of growth (expressed as a decimal)
i = initial yield
e = equated yield
n = number of years in rent review pattern.

Thus

$$(1 + g)^5 = \frac{\text{YP in perpetuity at 5 per cent} - \text{YP 5 years at 12 per cent}}{\text{YP in perpetuity at 5 per cent} \times \text{PV of £1 in 5 years at 12 per cent}}$$

$$(1 + g)^5 = \frac{20 - 3.605}{20 \times 0.567} = \frac{16.395}{11.340} = 1.446$$

$$\sqrt[5]{1.446} = 1 + g \qquad = 1.0765$$

$$g \qquad = 0.0765$$

The implied rate of rental growth is therefore 7.65 per cent per annum.
The inflation risk free yield (IRFY) is determined using the formula:

$$\frac{1 + e}{1 + g} - 1 = i$$

where

e = equated yield
g = annual growth rate } all expressed as decimals
i = IRFY

$$\frac{1.12}{1.0765} - 1 = 0.0404$$

and IRFY = 4.04 per cent.

Although this figure is not rounded off, it would be acceptable, in an examination, to round off the IRFY to 4 per cent, for ease of reference to valuation tables.

The current rental value on the basis of 5 year reviews may now be converted on to the basis of a rent fixed for 21 years, using the formula:

$$X \times \frac{\text{YP } n \text{ years at } e}{\text{YP } n \text{ years at } i} \times \frac{\text{YP } N \text{ years at } i}{\text{YP } N \text{ years at } e} = y$$

where

X = known rental value on existing review pattern
n = number of years in existing review pattern
e = equated yield
i = IRFY
N = number of years in new review pattern
Y = rental value on new review pattern.

Thus, the rental value on a 21 year review pattern =

$$£22\,000 \times \frac{\text{YP 5 years at 12 per cent}}{\text{YP 5 years at 4.04 per cent}} \times \frac{\text{YP 21 years at 4.04 per cent}}{\text{YP 21 years at 12 per cent}}$$

$$= £22\,000 \times \frac{3.605}{4.447} \times \frac{13.978}{7.562}$$

$$= £32\,966$$

say £32 965 pa

These calculations are 3 years prior to the rent review and, at this stage, no account is taken of rental growth during that time.

The rental of £32 965 pa is almost a 50 per cent uplift from £22 000 per annum and it is distinctly possible that the freeholder will have difficulty in persuading the head leaseholder to accept such an increase, even though the rent will be fixed for 21 years.

An alternative approach might be analysis to discover the magnitude of uplift, for each year's difference in rent review patterns, that is being accepted in the market.

Assuming that this difference was found to be between 1 per cent and 2 per cent for each year:

1 per cent would result in 21 years − 5 years

$$= 16 \text{ years at } 1 \text{ per cent}$$

$$= 16 \text{ per cent uplift}$$

2 per cent would result in a 32 per cent uplift.

For the purposes of this question, it will be assumed that the uplift is agreed at $1\frac{1}{2}$ per cent per annum and that the outcome of negotiations between freeholder and head leaseholder would be a rent of £27 300 pa for the last 21 years of the head lease.

$$£22\,000 + 24 \text{ per cent of } £22\,000$$

$$= £22\,000 + £5\,280 = £27\,280$$

say £27 300 pa

(ii) This part of the question requires consideration of the valuation of the head leasehold interest using a traditional approach, therefore it is necessary to assume a remunerative yield, an accumulative yield and a rate of income tax.

Assuming a remunerative yield of 6 per cent, income tax at 40p in the £ and an annual sinking fund at 3 per cent, a traditional valuation would appear as follows:

Rent received		£20 000 pa	
less Rent paid		£ 7 500 pa	
Net income		£12 500 pa	
YP 3 years at 6 per cent and 3 per cent (tax 40 per cent)		1.669	£20 863
Rent received		£20 000 pa	
less Rent paid		£27 300 pa	
Net income		−£ 7 300 pa	
YP 1 year at 6 per cent and 3 per cent (tax 40 per cent)	0.579		
x PV of £1 in 3 years at 6 per cent	0.84	0.486	NIL
Rent received		£22 000 pa	
less Rent paid		£27 300 pa	
Net income		−£ 5 300 pa	
YP 20 years at 6 per cent and 3 per cent (tax 40 per cent)	8.195		
x PV of £1 in 4 years at 6 per cent	0.792	6.490	NIL
	Capital value		£20 863

A traditional valuation shows a profit rent or net income for only 3 years, after which the profit rent becomes negative, strictly producing a negative value. A nil capital value will usually be adopted, unless the valuer ascribes some intuitive value to it.

The method makes no allowance for the different structures of the two leases — the head lease rent is fixed for 21 years after review in 3 years' time, whereas the sublease rent is to be reviewed every 5 years and the £22 000 per annum will not, therefore, be fixed. No account is taken of the fact that the rent received by the head leaseholder will be subject to these reviews — it appears that both rent paid and rent received are fixed, causing the head leasehold interest to be under-valued.

(iii) Although it is required that this part of the question is answered using either a discounted cash flow or a real value approach, both will be shown here, to demonstrate the methods.

Both approaches overcome the difficulties encountered when applying the traditional approach, by treating the valuation of the two leases entirely separately. The capital value of the head lease rent is deducted from the capital value of the sublease rent. The two rents are capitalised separately, so that the growth differential may be accounted for — the rent received by the head lessee is on the basis of 5 year reviews and the rent paid is fixed.

Valuation of the head leasehold interest using a discounted cash flow at equated yield

An equated yield of 12 per cent was used for analysis of the freehold initial yield, but, for the valuation of the head leasehold interest, a higher equated yield is required to allow for the extra risks attached to leasehold interests. In this valuation, an equated yield of 18 per cent will be adopted.

Capital value of rent received

Year	Rent £	Amount of £1 at 7.65 per cent[a]	Inflated rent £	YP at 18 per cent[b]	PV of £1 at 18 per cent[b]	Present value £	
1–4	20 000	1	20 000	2.690	1	53 800	
5–9	22 000[c]	1.343	29 546	3.127	0.516	47 673	
10–14	29 546	1.446	42 724	3.127	0.226	30 193	
15–19	42 724	1.446	61 779	3.127	0.099	19 125	
20–24	61 779	1.446	89 332	3.127	0.043	12 012	162 803

Less
Capital value of rent paid:

| 1–3 | 7 500 | 1 | 7 500 | 2.174 | 1 | 16 305 | |
| 4–24 | 27 300[d] | 1.2475 | 34 057 | 5.384 | 0.609 | 111 668 | 127 973 |

Capital value of head leasehold interest — £ 34 830

say £35 000

[a]Rent received grows at 7.65 per cent per annum, realiseable every 5 years after year 4.
[b]Equated yield.
[c]Review to current rental value on the basis of reviews every 5 years.
[d]Review to current rental value on the basis of a rent fixed for 21 years.

Valuation of head leasehold interest using a real value approach

First of all, it is necessary to recalculate the inflation risk free yield for the head leasehold interest, using an equated yield of 18 per cent.

$$\frac{1+e}{1+g} - 1 = \frac{1.18}{1.0765} - 1 = 0.0961$$

IRFY = 9.61 per cent

As with the discounted cash flow approach, the method employed is to deduct the capital value of the rent paid from the capital value of the rent received, so that the difference in growth potential of the two lease rents may be taken into account.

Capital value of rent received:

Rent received	£20 000 pa	
YP 4 years at 18 per cent [see note 1]	2.690	£53 800
Reversion to	£22 000 pa	

$$\text{YP 20 years at 9.61 per cent} \times \frac{\text{YP 5 years at 18 per cent*}}{\text{YP 5 years at 9.61 per cent}}$$

*[see note 2]

$$= 8.745 \times \frac{3.127}{3.829} \qquad\qquad = 7.142$$

x PV of £1 in 4 years at 9.61 per cent	0.693	4.949	£108 878	£162 678
[see note 3]				

less
Capital value of rent paid:

Rent paid		£ 7 500 pa		
YP 3 years at 18 per cent [see note 4]		2.174	£16 305	
Reversion to		£27 300 pa		
YP 21 years at 18 per cent [see note 5]	5.384			
x PV of £1 in 3 years at 9.61 per cent	0.759	4.086	£111 548	£127 853
[see note 6]				
Capital value of head leasehold interest				£ 34 825

    say £35 000

Notes
1: Rent fixed for 4 years, therefore valued at equated yield.
2: Three YPs reflect the fact that the income lasts for 20 years, also that rental growth is at 7.65 per cent per annum, realised every 5 years to produce an overall yield of 18 per cent.
3: Deferred at inflation risk free yield to reflect growth in rental value during the first 4 years.
4: Rent fixed for 3 years therefore valued at equated yield.
5: Rent fixed for 18 years therefore valued at equated yield.
6: Deferred at inflation risk free yield to reflect growth in rental value during the first 3 years.

The capital value of £35 000 for the head leasehold interest is its value single rate and gross of tax. To obtain a dual rate, tax adjusted value, Pannell's adjustment may be applied.

Assuming tax at 40 per cent and an annual sinking fund of 3 per cent, the value will be

$$£35\,000 \times \frac{\text{YP 24 years at 18 per cent and 3 per cent (tax at 40 per cent)}}{\text{YP 24 years at 18 per cent}}$$

$$= £35\,000 \times \frac{4.378}{5.451} = £28\,110$$

say £28 000

Question 9.2

Valuation of freehold interest

All the required information is available to produce a valuation of the freehold interest using conventional techniques. However, in order to apply a real value approach, it is necessary to determine the equated yield, the rate of rental growth per annum and the inflation risk free yield.

In order to estimate the equated yield, it is acceptable to add 2 per cent to the yield from long dated gilt edged stock.

If no information is provided in the question, as in this instance, an assumption should be made, and a clear statement made regarding what that assumption is. In this case, it will be assumed that gilts are yielding 10 per cent and an equated yield of 12 per cent will be adopted.

The rate of rental growth implied by a rack rented capitalisation rate of 6.5 per cent, may be calculated using the formula:

$$(1 + g)^n = \frac{\text{YP in perpetuity at } i - \text{YP } n \text{ years at } e}{\text{YP in perpetuity at } i \times \text{PV of £1 in } n \text{ years at } e}$$
$$[\text{see Question 9.1}]$$

Thus

$$(1 + g)^5 = \frac{\text{YP in perpetuity at 6.5 per cent} - \text{YP 5 years at 12 per cent}}{\text{YP in perpetuity at 6.5 per cent} \times \text{PV of £1 in 5 years at 12 per cent}}$$

$$(1 + g)^5 = \frac{15.385 - 3.605}{15.385 \times 0.567} = \frac{11.780}{8.723} = 1.350$$

$$\sqrt[5]{1.350} = 1 + g = 1.06185$$

$$g = 0.06185$$

Implied rate of rental growth is therefore 6.185 per cent per annum.

The inflation risk free yield (IRFY) may be determined using the formula:

$$\frac{1+e}{1+g} - 1 = i \quad \text{(see Question 9.1)}$$

$$\frac{1.12}{1.06185} - 1 = 0.0548$$

and IRFY = 5.48 per cent.

Although this has not been rounded off, it would be acceptable to do so in an examination, for ease of reference to valuation tables.

All the necessary information is now available to carry out a valuation of the freehold interest, using both conventional and real value techniques.

Valuation of freehold interest using a traditional approach

Rent received		£70 000 pa	
YP 3 years at 5.5 per cent [see note 1]		2.698	£ 188 860
Reversion to [see note 2]		£75 000 pa	
YP 15 years [see note 3] at 6 per cent [see note 4]	9.712		
x PV of £1 in 3 years at 6 per cent [see note 5]	0.840	8.158	£ 611 850
Reversion to [see note 6]		£95 000 pa	
YP in perpetuity at 6.5 per cent [see note 7]	15.385		
x PV of £1 in 18 years at 6.5 per cent [see note 8]	0.322	4.954	£ 470 630
	Capital value		£1 271 340
say		£1 270 000	

Notes

1: Traditional yield pattern. Rack rented capitalisation rate, from comparables, is 6.5 per cent, term income capitalised at 1 per cent less, since it is considered to be more secure than reversionary income.

2: Reversion to rental value excluding improvements. It is only 7 years since the improvements were carried out by the tenant, therefore their value cannot be reflected in the rent at this stage ('21 year rule', *Landlord and Tenant Act 1954* and *Law of Property Act 1969*).

3: The present lease has expired and the landlord will want the new lease to be of such length that it terminates as soon as possible after 21 years have elapsed since the tenant carried out the improvements, at which point, the value of these improvements may be reflected in the rent. However, from the tenant's view-point, it will be desirable to obtain a lease of the longest possible duration, thus extending, for as long as possible, the period during which the value of the improvements is disregarded. The final agreement on the length of the new lease will depend upon the relative bargaining positions of the two parties. If referred to court, the maximum term that may be granted is 14 years and a 15 year lease has been adopted in this case. A student answering the question could make any reasonable assumption, substantiated by a brief explanation similar to that given above.

4: Capitalised at 0.5 per cent below the rack rented capitalisation rate. Although this is current rental value, it excludes the value of tenant's improvements and is assumed to be more secure than the current rental value when the value of the improvements is included.

5: Deferred at same yield as capitalisation rate.

6: At the end of the 15 year lease, it is 22 years since the tenant's improvements were carried out. The value of these improvements can now be reflected in the rent.

7: Yield of 6.5 per cent from comparable information. Reflects all risks attached to the income, rent on a 5 year review pattern and implies future growth in rental value.

8: Final reversion deferred at same yield as capitalisation rate.

Valuation of freehold interest using a real value approach

Rent received	£70 000 pa	
YP 3 years at 12 per cent [see note 1]	2.402	£ 168 140
Reversion to [see note 2]	£75 000 pa	

YP 15 years [see note 3] at 5.48 per cent $\times \dfrac{\text{YP 5 years at 12 per cent*}}{\text{YP 5 years at 5.48 per cent}}$

*[see note 4]

$$= 10.051 \times \frac{3.605}{4.273} = 8.480$$

× PV of £1 in 3 years at 5.48 per cent [see note 5]	0.852	7.225	£ 541 875
Reversion to [see note 6]		£95 000 pa	
YP in perpetuity at 6.5 per cent [see note 7]		15.385	
× PV of £1 in 18 years at 5.48 per cent [see note 8]		0.383	5.892	£ 559 740
	Capital value		£1 269 755
say			£1 270 000

Notes

1: Rent fixed for 3 years. No growth in income, therefore valued at equated yield.

2: See note 2, freehold interest, traditional approach.

3: See note 3, freehold interest, traditional approach.

4: Three YPs reflects the fact that the income is of 15 years' duration, also that rental growth is 6.185 per cent per annum, realised every 5 years to give an overall yield of 12 per cent.

5: Deferred at inflation risk free yield to reflect growth in the £75 000 per annum during the first 3 years.

6: See note 6, freehold interest, traditional approach.

7: Yield of 6.5 per cent from comparable information. Reflects rental growth at 6.185 per cent annum, realised every 5 years to give an overall yield of 12 per cent. Alternatively, the 3 YPs may be used:

$$\text{YP in perpetuity at 5.48 per cent} \times \frac{\text{YP 5 years at 12 per cent}}{\text{YP 5 years at 5.48 per cent}}$$

This reflects the same facts, although more explicitly, and will result in the same valuation.

8: Deferred at inflation risk free yield to reflect growth in the £95 000 per annum during the 18 years before reversion.

Valuation of leasehold interest

As no information is given in the question, an assumption must be made regarding the rate of income tax and the rate of interest available on the annual sinking fund. In this case, it will be assumed that income tax is at 40p in the £ and that an annual sinking fund is available at 3 per cent.

Valuation of head leasehold interest using a traditional approach

Current rental value [see note 1]	£95 000 pa		
less Rent paid	£70 000 pa		
Profit rent	£25 000 pa		
YP 3 years at 6.5 per cent and 3 per cent			
(tax at 40 per cent) [see note 2]	1.655		£ 41 375
Current rental value [see note 3]	£95 000 pa		
less Rent paid	£75 000 pa		
Profit rent	£20 000 pa		
YP 15 years [see note 4] at 7.5 per cent			
[see note 5] and 3 per cent			
(tax at 40 per cent)	6.075		
× PV of £1 in 3 years at 7.5 per cent			
[see note 6]	0.805	4.890	£ 97 800
	Capital value		£139 175

say £140 000

Notes

1: Value of improvements reflected immediately in the value of the leasehold interest.

2: Traditional yield pattern. Remunerative yield 1 per cent above freehold yield.

3: See note 2, freehold interest, traditional approach.

4: See note 3, freehold interest, traditional approach.

5: Traditional yield pattern. Remunerative yield 1 per cent above freehold yield.

6: Deferred at same rate as remunerative yield.

Valuation of head leasehold interest using a real value approach

An equated yield of 12 per cent was adopted in the valuation of the freehold interest, but, for the valuation of the head leasehold interest, a higher equated yield is necessary, to allow for the extra risks associated with leasehold interests. In this valuation, an equated yield of 18 per cent will be adopted.

The inflation risk free yield must also be recalculated on this basis.

$$\frac{1+e}{1+g} - 1 = \frac{1.18}{1.06185} - 1 = 0.1113$$

IRFY = 11.13 per cent

The method employed by the real value approach is to value the rent paid and the rent received (or notionally received) separately, so that the different growth prospects of the two rents is taken into account. The capital value of the rent paid is then deducted from the capital value of the rent received, to give the capital value of the leasehold interest.

Capital value of rent received:

Rent received [see note 1]　　　　　　　　　　　£95 000 pa

YP 18 years [see note 2] at 11.13 per cent × $\dfrac{\text{YP 5 years at 18 per cent}}{\text{YP 5 years at 11.13 per cent}}$

[see note 3]

$$= 7.640 \times \dfrac{3.127}{3.684} = \quad 6.485 \qquad £616\,075$$

less

Capital value of rent paid:

Rent paid　　　　　　　　　　　　　　　£70 000 pa
YP 3 years at 18 per cent [see note 4]　　2.174　　£152 180

Reversion to　　　　　　　　　　　　　£75 000 pa

YP 15 years [see note 5] at 11.13 per cent × $\dfrac{\text{YP 5 years 18 per cent}}{\text{YP 5 years at 11.13 per cent}}$

[see note 6]

$$= 7.140 \times \dfrac{3.127}{3.684} = 6.060$$

× PV of £1 in 3 years at 11.13 per cent　0.729　4.418　£331 350　£483 530
[see note 7]

　　　　　　　　　　　　　Capital value　　　　　　　　£132 545

Notes

1: See note 1 leasehold interest, traditional approach.
2: See note 3, freehold interest, traditional approach.
3: Three YPs reflects the fact that the income is of 18 years' duration, also that rental growth is 6.185 per cent per annum, realised every 5 years to give an overall yield of 18 per cent.
4: See note 2, freehold interest, traditional approach.
5: See note 3, freehold interest, traditional approach.
6: Three YPs reflects rental growth of 6.185 per cent per annum, realised every 5 years for 15 years, giving an overall yield of 18 per cent.
7: Deferred at inflation risk free yield to reflect growth in the rental value during the 3 years before reversion.

Similarities and differences between the two approaches

This part of the answer is best dealt with by considering freehold and leasehold interests separately and then comparing the two valuation approaches stage by stage, where possible.

Freehold interest

Valuation of the term income

The first item to note is that, in the traditional approach, a yield of 1 per cent below the rack rented capitalisation rate is used to value the term income. The reasoning behind this is that the income is below full rental value and is considered to be secure — the tenant has a profit rent and is therefore less likely to default. Before inflation was a problem, this yield would have been logical, since it would have been reasonably in line with the yield from gilt edged stock, thus treating a fixed income from property in the same way as other fixed income investments. However, in an inflationary economy, the use of a low yield (5.5 per cent in this instance) is illogical. A low, growth-implicit yield is being used in a situation where there is no growth — the rent of £70 000 per annum is fixed for the next 3 years. The result is an over-valuation of the term income.

In contrast, the real value approach uses the equated yield in the valuation of the term income. It is accepted practice (although arguable) to use a yield of 2 per cent above the yield on long dated gilt edged securities. Thus a fixed income from a property investment is treated on the same basis as other fixed income investments — adopting the original logic of the traditional approach. The 2 per cent differential is to allow for the extra risks of investing in property compared with other investments and is the yield gap that existed between prime property investments and gilts, prior to the recognition of inflation as a serious consideration. It is argued that, since 2 per cent was the differential before inflation was a problem, then 2 per cent must still be the true basic difference in yield. Until further research produces a more satisfactory basis, this approach is acceptable, since it is reasonable and logical. In the valuation in question, the £70 000 per annum is thus correctly treated as remaining fixed for the next 3 years.

Valuation of the middle term

In both approaches, it is assumed that the middle term is for a period of 15 years (or whatever length of lease is adopted in the answer) and that the rent received is the current rental value disregarding the tenant's improvements.

The traditional approach values this income at a yield somewhere between the yield used to value the first term and that used to value the final reversionary income. The same argument is used as that applied to the first-term valuation, that is, security of income because the tenant has a profit rent. The use of this low yield for the middle-term valuation is more logical than in the first-term valuation, because the income of £75 000 per annum is not fixed for the whole 15 years, but is on a 5 year review pattern.

Thus, reflection of some growth in this income is correct, but, here again, the yield used, being less than the rack rented capitalisation rate, implies more growth than actually occurs. Some degree of over-valuation is again the result.

In the application of the real value approach, it should be explained that the valuer is able to take account of the actual length of the middle term and the review pattern, using the formula:

$$\text{YP } N \text{ years at } i \times \frac{\text{YP } n \text{ years at } e}{\text{YP } n \text{ years at } i}$$

where

N = length of term
i = inflation risk free yield
n = number of years in rent review pattern
e = equated yield.

This will reflect the correct amount of growth in income over the 15 years, with a 5 year review pattern, to give the overall required yield (the equated yield). The formula can be used to deal with any term length and any review pattern.

Comment should be made on the yield used in the two approaches to defer the value of the middle term.

The traditional method employs the same yield as that used to value the income. Rental value in the middle term is taken to be the current rental value at the time of valuation, since, when the traditional method was devised, in pre-inflationary times, it was assumed that rents would not rise. Past experience had been of both increases and decreases in rental values, therefore it became accepted to assume that they would remain static.

Although the yield used to defer the value of the middle term is growth-implicit, it does not allow sufficiently for growth in rental value during the 3 years of the first term. Since the first term is short, this under-allowance for growth is not sufficient to compensate for the initial over-valuation of the income.

Reversion in the middle term is also to current (unimproved) rental value in the real value approach, but in this case, the capital value is deferred at a lower yield — the inflation risk free yield — to reflect growth correctly in the rental value during the 3 years of the first term. Alternatively, the valuer may be explicit regarding the rental growth. The income for the middle term would be shown as the estimated unimproved rental value in 3 years' time, with rental growth at 6.185 per cent per annum. If this is done, the middle-term value is deferred at the equated yield — a yield less than this would incorrectly imply further growth in rental value.

Middle term — alternative real value approach

Rent received	£ 75 000 pa
× Amount of £1 in 3 years at 6.185 per cent	1.197
[see note 1]	
Estimated rental value in 3 years' time	£ 89 775 pa

$$\text{YP 15 years at 5.48 per cent} \times \frac{\text{YP 5 years at 12 per cent}}{\text{YP 5 years at 5.48 per cent}}$$

$$= 10.051 \times \frac{3.605}{4.273} = 8.480$$

× PV of £1 in 3 years at 12 per cent
[see note 2]

	0.712	6.038
Capital value of middle term		£542 061

Notes

1: Growth in rental value during the 3 years of the first term.

2: Deferred at equated yield. Rental growth has been shown explicitly, therefore there is no need, indeed it would be incorrect, to reflect it in the yield.

Valuation of the reversionary income

It should firstly be noted that, in both cases, the valuation into perpetuity is achieved using the rack rented capitalisation rate, obtained from the transaction involving a similar property. In practice, it is always desirable, where possible, to work on information from more than one comparable, since there is always a danger that an isolated transaction may be unique for some reason that in turn affected the yield. However, in examination questions, it is usually safe to assume, unless the wording of the question implies otherwise, that any comparables given conform to the norm for the type of property under consideration.

In this case, since the income from the comparable property and that from the subject property are on the basis of 5 year reviews, the 6.5 per cent yield may be applied directly in the valuation. No manipulation of the yield is required. In both the traditional and real value approaches, the YP in perpetuity at 6.5 per cent reflects a 5 year review pattern and future rental growth of 6.185 per cent per annum. Reversion in both cases is to current rental value reflecting the value of the tenant's improvements.

As in the middle term, the two methods differ in the yield used to defer the reversionary value. The traditional method employs the same yield of 6.5 per cent used to value the income, for the reasons previously explained when considering the middle term valuation. This causes an under-valuation of the reversion. The over-valuation of the term incomes and under-valuation of the reversion, in the traditional approach, do tend to compensate, but the extent to which this occurs depends upon the length of time to final reversion and the relative values of term and reversion.

Reversion is also to current rental value in the real value approach, but the reversionary value is deferred at the inflation risk free yield, to reflect growth in the rental value during the 18 years before reversion. As with the middle term, the valuer may alternatively show growth explicitly in the rental value, then deferring at the equated yield.

Leasehold interest

The student will have difficulty in comparing the approaches of the two methods to valuation of the leasehold interest, since they are vastly different.

The traditional method employs the concept of a profit rent, which is valued in term and reversion format. In both the term and reversion, the leaseholder is able to sublet the property at its improved rental value of £95 000 per annum. During the term, he pays a rent of £70 000 per annum, thus enjoying a profit rent of £25 000 per annum. This is capitalised using a dual rate, tax adjusted years' purchase. A reasonable sinking fund rate of 3 per cent has been assumed, with income tax being paid at 40 per cent.

The remunerative yield used in the valuation of the profit rent for the term is conventionally taken to be approximately 1 per cent more than the yield adopted to value the income in the corresponding part of the freehold valuation. This is to allow for the extra risks involved in a leasehold investment compared to a freehold.

The same process is repeated in the reversionary valuation, where the profit rent falls to £20 000 per annum. Deferment is carried out at the same yield as that used in the valuation, following the same arguments as those applicable to the freehold valuation.

However, there are various problems associated with the application of this traditional approach.

The first of these is the mathematical error which occurs in the valuation of a varying profit rent. For a consideration of this aspect, the reader is referred to chapter 1 of *Applied Valuation* by Diane Butler (Macmillan, 1987). Reference will also be found in that chapter to the question of the validity of an annual sinking fund assumption and the low rates of interest conventionally adopted.

A further problem is the concept of valuing a profit rent, which leads to the valuation of an income derived from two incomes on differing bases. For example, the profit rent for the term in the example question, is derived as follows:

Current rental value		£95 000 pa
less Rent paid		£70 000 pa
	Profit rent	£25 000 pa

However, the current rental value is on the basis of a rent reviewed every 5 years, whereas the rent paid is fixed. The £25 000 per annum is thus a 'mixture' of these two figures on different bases, which is valued at a growth implicit yield, derived from freehold comparables. This leads on to a further criticism of the traditional method — basing the yield for a leasehold valuation upon freehold yields from similar properties, on the arbitrary basis of the addition of approximately 1 per cent. Leasehold interests are not comparable with freehold interests and, indeed, are very rarely comparable with other leasehold interests. For instance, lengths of unexpired leases differ and gearing varies. Why then should they be valued at a yield derived from freehold interests?

The real value method attempts to overcome these difficulties by using a completely contrasting approach, abandoning the concept of profit rent. The valuation of the rent received (or the rent that could be received) is treated separately from the valuation of the rent paid, so that the individual growth prospects and review patterns of the two incomes may be correctly reflected.

The rent received, £95 000 per annum, is of 18 years' duration and is reviewable every 5 years. Valuation of this income is achieved using the '3 YP' formula. The equated yield adopted in this case will be some 6 to 8 per cent more than the yield obtainable from gilts, to allow for the extra risks involved in leasehold property investments. Thus, in the valuation of the £95 000 per annum, account will be taken of the fact that the income will be received for 18 years, rental growth is 6.185 per cent per annum and is realised on review every 5 years. This will produce an overall yield equal to the required equated yield. The valuation may equally be achieved using a discounted cash flow approach, using the equated yield, since this is the basis of the real value method.

The rent paid is valued in two stages. For the first 3 years, the rent paid is fixed and is valued at the equated yield to reflect this. After 3 years, the rent rises to £75 000 per annum, which is reviewable every 5 years and the '3 YP' formula is used in this part of the valuation. Deferment is undertaken using the inflation risk free yield, to reflect growth in the £75 000 per annum during the 3 years before it becomes payable.

Having derived the capital value of the rent paid, it is deducted from the capital value of the rent received, giving the capital value of the leasehold interest.

In conclusion, it may be useful to note that the real value approach is an attempt to overcome many of the criticisms aimed at the traditional approach in an inflationary economy. However, choice of equated yield remains somewhat unsatisfactory and further research is needed to add stringency to this element of the valuation.

Question 9.3

Whichever equated yield approach is adopted, there are several variables which must be determined before the required valuations can be performed. From comparable information, it is necessary to calculate the all risks yield, the Zone A rental value per m^2 and the annual rate of rental growth. Since an equated yield approach to the valuations is required, the equated yield must also be determined — or estimated in this case, since no information is given in the question.

If the chosen equated yield approach is by discounted cash flow, then no further analysis is required. If, however, the student decides to adopt a real value approach, then the inflation risk free yield must also be calculated.

The initial stages of this answer therefore involve analysis of the given information.

The initial (all risks) yield on the basis of 3 year reviews may be discovered using the information from the transaction involving the nearby similar shop, since this is a recent letting, followed by sale of the freehold interest.

$$\text{Initial yield} = \frac{£21\,750}{£435\,000} \times 100 = 5 \text{ per cent}$$

The next step is analysis of the rent achieved on letting the similar property, to discover the Zone A rental value per m^2.

Two 5 m zones and a remainder will be adopted.

Let the Zone A rental value per $m^2 = £X$

Zone A $= 5 \text{ m} \times 5 \text{ m} \times \quad X = 25X$
Zone B $= 5 \text{ m} \times 5 \text{ m} \times \frac{1}{2} X = 12.5X$
Remainder $= 5 \text{ m} \times 5 \text{ m} \times \frac{1}{4} X = \underline{6.25X}$

Rental value $\underline{43.75X}$

$$43.75X = £21\,750$$
$$\text{and} \quad X = £497 \text{ per } m^2$$

The initial yield from the comparable property may now be analysed to determine the annual rate of rental growth implied by the acceptance of a low initial yield of 5 per cent. For this purpose, it is necessary to know the equated yield and, since no indication is given in the question, a reasonable assumption should be made. In this instance, it will be assumed that gilts are yielding 10 per cent and a freehold equated yield of 12 per cent will be adopted (see Question 9.2).

Implied growth rate is derived using the formula:

$$(1 + g)^n = \frac{\text{YP in perpetuity at } i - \text{YP } n \text{ years at } e}{\text{YP in perpetuity at } i \times \text{PV of £1 in } n \text{ years at } e}$$

(see Question **9.1**)

$$(1 + g)^3 = \frac{\text{YP in perpetuity at 5 per cent} - \text{YP 3 years at 12 per cent}}{\text{YP in perpetuity at 5 per cent} \times \text{PV of £1 in 3 years at 12 per cent}}$$

$$(1 + g)^3 = \frac{20 - 2.402}{20 \times 0.712} = \frac{17.598}{14.240} = 1.236$$

$$\sqrt[3]{1.236} = 1 + g = 1.0731$$

$$g = 0.0731$$

The implied rate of rental growth is therefore 7.31 per cent per annum.

The inflation risk free yield (IRFY) is determined using the formula:

$$\frac{1 + e}{1 + g} - 1 = i \quad \text{(see Question 9.1)}$$

$$\frac{1.12}{1.0731} - 1 = 0.0437$$

and IRFY = 4.37 per cent.

This figure has not been rounded off, but it would be acceptable to do so in an examination, for ease of reference to the valuation tables.

The Zone A rental value per m^2, obtained from the letting of the similar property, is now used to determine the current rental value of the subject property. Once again, 5.m zones must be used, so that the properties are compared on the same basis.

```
Zone A      = 6 m ×  5 m × £497      = £14 910
Zone B      = 6 m ×  5 m × £248.5    = £  7 455
Remainder = 6 m × 10 m × £124.25   = £  7 455
                                      ─────────
                                        £29 820
                                      ─────────
```

Therefore the current rental value of the subject property, on full repairing and insuring terms and on the basis of 3 year reviews is £29 820 pa.

All the information is now available to carry out the necessary valuation of the freehold interest. Although only one approach is required, the valuation will be done using both a discounted cash flow at equated yield and a real value approach, in order to demonstrate the methods.

Valuation of freehold interest using a discounted cash flow at equated yield approach

Year	Rent £	Amount of £1[a] at 7.31 per cent	Inflated rent £	YP at 12 per cent[b]	PV at 12 per cent[c]	Present value £
1–26	2 500	1	2 500	7.896	1	19 740
27–29	29 820[d]	6.261	186 703	2.402	0.052	23 320
30–32	186 703	1.236	230 765	2.402	0.037	20 509
33+[e]	230 765	1.236	285 226			
YP in perpetuity at 5 per cent[f]		20				
x PV of £1 32 years at 12 per cent[g]		0.027	0.54			£154 022
			Capital value			£217 591
			say £215 000			

[a]Rental growth at 7.31 per cent per annum.
[b]Valued at equated yield.
[c]Deferred at equated yield.
[d]The current lease is now ended and the property may be let at its current rental value. It is assumed that this will be on the basis of 3 year rent reviews.
[e]It is acceptable to discontinue a discounted cash flow valuation at around 30 years. The review at the end of year 32 is a convenient point to do this.
[f]Income is now valued into perpetuity at the rack rented capitalisation rate from the transaction involving the similar property. This rate reflects a 3 year review pattern and rental growth of 7.31 per cent per annum, to give an overall yield of 12 per cent.
[g]Deferred at equated yield since rental growth up to this point has been shown explicitly in the rent.

Valuation of freehold interest using a real value approach

Rent received		£ 2 500 pa	
YP 26 years at 12 per cent		7.896	£ 19 740
[see note 1]			
Reversion to [see note 2]		£ 29 820 pa	
YP in perpetuity at 5 per cent [see note 3]	20		
x PV of £1 in 26 years at 4.37 per cent [see note 4]	0.329	6.58	£196 216
	Capital value		£215 956
	say	£215 000	

Notes

1: Fixed rent for 26 years. No growth in income, therefore valued at equated yield.

2: Estimated current rental value on full repairing and insuring terms with 3 yearly reviews.

3: All risks yield from comparable information. Reflects rental growth at 7.31 per cent per annum and rent reviews every 3 years, to give an overall yield of 12 per cent.

4: Deferred at inflation risk free yield to reflect growth in the £29 820 during the 26 years before reversion.

Although the head lessee sublet the subject property 7 years ago on the basis of 5 year reviews, the most recent evidence has been adopted in the valuation and it has been assumed that the subject property would re-let on a 3 year review pattern.

However, the following valuation is included should the student decide to adopt the assumption of a 5 year review pattern.

Initially, the current rental value on the basis of 3 year reviews must be converted on to the basis of 5 year reviews using the formula:

$$X \times \frac{\text{Years' purchase } n \text{ years at } e}{\text{Years' purchase } n \text{ years at } i} \times \frac{\text{Years' purchase } N \text{ years at } i}{\text{Year's purchase } N \text{ years at } e} = y$$

(see Question 9.1)

Thus, the rental value on a 5 year review pattern

$$= £29\,820 \times \frac{\text{YP 3 years at 12 per cent}}{\text{YP 3 years at 4.37 per cent}} \times \frac{\text{YP 5 years at 4.37 per cent}}{\text{YP 5 years at 12 per cent}}$$

$$= £29\,820 \times \frac{2.402}{2.756} \times \frac{4.406}{3.605} = £31\,764$$

say £31 765 pa

Valuation

Rent received	£ 2 500 pa	
YP 26 years at 12 per cent	7.896	£19 740
Reversion to [see note 1]	£31 765 pa	

$$\text{YP in perpetuity at 4.37 per cent} \times \frac{\text{YP 5 years at 12 per cent}}{\text{YP 5 years at 4.37 per cent}}$$

[see note 2]

$$= 22.883 \times \frac{3.605}{4.406} = 18.723$$

× PV of £1 in 26 years at 4.37 per cent	0.329	6.16	£195 672
		Capital value	£215 412

say £215 000

Notes

1: Estimated current rental value on the basis of 5 year reviews.

2: Three YPs reflects 5 year review pattern and growth at 7.31 per cent to give an overall yield of 12 per cent.

Thus, it can be seen that whether the current rental value is assumed to be on the basis of 3 year or 5 year reviews, it makes no difference to the capital value of the freehold interest — the YP is adjusted to take account of the review pattern.

Valuation of head leasehold interest using a discounted cash flow at equated yield approach

An equated yield of 12 per cent was adopted in the valuation of the freehold interest, but a higher yield is necessary to value the leasehold interest (see Question 9.2). In this case, an equated yield of 18 per cent will be adopted.

Year	Rent £	Amount of £1 at 7.31 per cent[b]	Inflated rent £	YP at 18 per cent[c]	PV at 18 per cent[d]	Present value £
Capital value of rent received[a]						
1–3	27 000[e]	1	27 000	2.174	1	58 698
4–8	31 765[f]	1.236	39 262	3.127	0.609	74 768
9–13	39 262	1.423	55 870	3.127	0.266	46 472
14–16	29 820[g]	2.502	74 610	2.174	0.116	18 815
17–19	74 610	1.236	92 218	2.174	0.071	14 234
20–22	92 218	1.236	113 981	2.174	0.043	10 655
23–25	113 981	1.236	140 881	2.174	0.026	7 963
26	140 881	1.236	174 129	0.847	0.016	2 360
						233 965[h]
less capital value of rent paid[a]						
1–26	2 500	1	2 500	5.480	1	13 700[i]
			Capital value of head leasehold interest			£220 265[j]

say £220 000

[a]Rent received and rent paid by head leaseholder are valued separately so that the differing growth prospects of the two incomes may be taken into account.
[b]Rental growth at 7.31 per cent per annum.
[c]Fixed income valued at equated yield.
[d]Deferred at equated yield.
[e]Rent fixed for next 3 years until review.
[f]Reversion to current rental value on the basis of 5 year rent reviews.
[g]Reversion to current rental value on the assumption that rent will be reviewed every 3 years. Alternatively, it may be assumed that the property will be re-let on the basis of 5 year reviews.
[h]Capital value of rent received by head leaseholder.
[i]Capital value of rent paid by head leaseholder.
[j]Value of rent paid is deducted from value of rent received to give capital value of head leasehold interest.

Capital value of head leasehold interest using a real value approach

Before commencing the valuation, the inflation risk free yield must be re-calculated using an 18 per cent equated yield.

$$\frac{1+e}{1+g} - 1 = \frac{1.18}{1.0731} - 1 = 0.0996$$

and IRFY = 9.96 per cent.

It would be acceptable to round this off to say 10 per cent in an examination.

Capital value of rent received:

[see note 1]		
Rent received	£27 000 pa	
YP 3 years at 18 per cent [see note 2]	2.174	£58 698
Reversion to [see note 3]	£31 765	

$$\text{YP 10 years at 9.96 per cent} \times \frac{\text{YP 5 years at 18 per cent}}{\text{YP 5 years at 9.96 per cent}}$$
$$\text{[see note 4]}$$

$$= 6.155 \times \frac{3.127}{3.795} = 5.072$$

x PV of £1 in 3 years at 9.96 per cent [see note 5]	0.752	3.814	£121 152
Reversion to [see note 6]		£29 820	

$$\text{YP 13 years at 9.96 per cent} \times \frac{\text{YP 3 years at 18 per cent}}{\text{YP 3 years at 9.96 per cent}}$$
$$\text{[see note 7]}$$

$$= 7.118 \times \frac{2.174}{2.489} = 6.217$$

x PV of £1 in 13 years at 9.96 per cent [see note 8]	0.291	1.809	£ 53 944
		[see note 9]	£233 794

less capital value of rent paid:

[see note 1]			
Rent paid	£2 500 pa		
YP 26 years at 18 per cent [see note 10]	5.480		£ 13 700
		[see note 11]	
Capital value of head leasehold interest			£220 094
		[see note 12]	

say £220 000

Notes

1: See note[a], head leasehold interest, discounted cash flow approach.
2: Rent fixed for 3 years. No growth in income therefore valued at equated yield.
3: See note[f], head leasehold interest, discounted cash flow approach.
4: Three YPs reflect total length of unexpired term of sublease, rental growth of 7.31 per cent per annum, realised every 5 years on review, to give an overall yield of 18 per cent.
5: Deferred at inflation risk free yield to reflect growth in the £31 765 pa during the 3 years before reversion.
6: See note[g], head leasehold interest, discounted cash flow approach.
7: Three YPs reflect total length of unexpired term, rental growth of 7.31 per cent per annum realised every 3 years on review, to give an overall yield of 18 per cent.
8: Deferred at inflation risk free yield to reflect growth in the £29 820 pa during the 13 years before reversion.
9: Capital value of rent received by head leaseholder.
10: Rent fixed for 26 years. No growth in income, therefore valued at equated yield.
11: Capital value of rent paid by head leaseholder.
12: See note[j], head leasehold interest, discounted cash flow approach.

The following valuation illustrates the situation should it be decided to assume that the property will be re-let on the basis of 5 year rent reviews when the present sub-lease ends.

Capital value of rent received:

Rent received	£27 000 pa	
YP 3 years at 18 per cent	2.174	£ 58 698
Reversion to [see note 1]	£31 765 pa	

$$\text{YP 23 years at 9.96 per cent} \times \frac{\text{YP 5 years at 18 per cent}}{\text{YP 5 years at 9.96 per cent}}$$
[see note 2]

$$= 8.909 \times \frac{3.127}{3.795} = 7.341$$

x PV of £1 in 3 years at 9.96 per cent [see note 3]	0.752	5.52	£175 343
			£234 041

less *capital value of rent paid:*

Rent paid	£ 2 500 pa	
YP 26 years at 18 per cent	5.480	£ 13 700
Capital value of head leasehold interest		£220 341

say £220 000

Notes

1: Reversion to current rental value on the basis of reviews every 5 years.

2: Three YPs reflect total length of the unexpired term, rental growth of 7.31 per cent per annum and rent reviews every 5 years to give an overall yield of 18 per cent.

3: Deferred at inflation risk free yield to reflect growth in the £31 765 pa during the 3 years before reversion.

This again demonstrates that whether it is assumed that the head leaseholder re-lets the property on the basis of 3 year or 5 year reviews, it makes no difference to the value of the head leasehold interest.

Question 9.4

Information from the transaction involving the similar freehold property must first of all be analysed to determine the initial yield and the rate of annual rental growth implied by the acceptance of that initial yield. A further unknown variable is the equated yield and, if the real value approach is adopted, the inflation risk free yield must be calculated. No guidance is given in the question regarding the equated yield, therefore a reasonable assumption should be made. In this case, a freehold equated yield of 12 per cent will be adopted and leasehold equated yield of 18 per cent (see Question 9.2). The comparable property has recently been let and also the freehold interest sold. Using the information from these two transactions will enable derivation of the initial yield on the basis of 5 year rent reviews.

$$\text{Initial yield} = \frac{£42\,000}{£700\,000} \times 100 = 6 \text{ per cent}$$

The annual rate of rental growth implied by an initial yield of 6 per cent, may now be derived using the formula:

$$(1 + g)^n = \frac{\text{YP in perpetuity at } i - \text{YP } n \text{ years at } e}{\text{YP in perpetuity at } i \times \text{PV of £1 in } n \text{ years at } e}$$
$$\text{(see Question 9.1)}$$

$$(1 + g)^5 = \frac{\text{YP in perpetuity at 6 per cent} - \text{YP 5 years at 12 per cent}}{\text{YP in perpetuity at 6 per cent} \times \text{PV of £1 in 5 years at 12 per cent}}$$

$$(1 + g)^5 = \frac{16.667 - 3.605}{16.667 \times 0.567} = \frac{13.062}{9.450} = 1.382$$

$$\sqrt[5]{1.382} = 1 + g = 1.0668$$

$$g = 0.0668$$

The implied rate of rental growth is therefore 6.68 per cent per annum.

The inflation risk free yield may now be determined using the formula:

$$\frac{1+e}{1+g} - 1 = i \quad \text{(see Question 9.1)}$$

$$\frac{1.12}{1.0668} - 1 = 0.0499$$

and IRFY = 4.99 say 5 per cent.

Freehold interest

There are 23 years of the head lease unexpired, with a review to market value in 2 years' time. However, there are no further reviews and the rent will therefore be fixed for the last 21 years of the head lease. Because of this, the freeholder will require a higher rent than he would if the rent were to be reviewed every 5 years.

In order to estimate the rent required by the freeholder, constant rent theory may be used. This will enable conversion of the current rental value on a 5 year review pattern on to the basis of current rental value with no review for 21 years, using the formula:

$$X \times \frac{\text{YP } n \text{ years at } e}{\text{YP } n \text{ years at } i} \times \frac{\text{YP } N \text{ years at } i}{\text{YP } N \text{ years at } e} = y$$
(see Question 9.1)

Thus the rent on a 21 year review pattern

$$= £37\,500 \times \frac{\text{YP 5 years at 12 per cent}}{\text{YP 5 years at 5 per cent}} \times \frac{\text{YP 21 years at 5 per cent}}{\text{YP 21 years at 12 per cent}}$$

$$= £37\,500 \times \frac{3.605}{4.330} \times \frac{12.821}{7.562} = £52\,934$$

say £52 950 pa

This is an uplift of just over 41 per cent to compensate the freeholder for accepting a fixed rent for 21 years rather than having reviews every 5 years. It may prove difficult for the freeholder to persuade the head leaseholder to accept such a large increase. An alternative would be analysis of available information to determine the percentage uplift in rental value that is currently being agreed in the market for each years' difference in rent review patterns.

Assuming this increase is found to be in the region of 1.5 per cent, the resultant rent would be:

21 years − 5 years = 16 years
16 years at 1.5 per cent = 24 per cent
£37 500 + 24 per cent = £37 500 + £9 000 = £46 500 pa

For the purposes of this question, it will be assumed that the freeholder and head leaseholder agree a current rental, on the basis of a rent fixed for 21 years, of £46 500 per annum.

Although the question requires only one approach to the answer, both discounted cash flow and a real value approach will be demonstrated here.

Valuation of freehold interest using a discounted cash flow at equated yield approach

Year	Rent £	Amount of £1 at 6.68 per cent[a]	Inflated rent £	YP at 12 per cent[b]	PV at 12 per cent[c]	Present value £
1–2	2 000[d]	1	2 000	1.690	1	3 380
3–23	46 500[e]	1.138	52 917	7.562	0.797	318 926
24–28	37 500[f]	4.425	165 937	3.605	0.074	44 267
29–33	165 937	1.382	229 325	3.605	0.042	34 722
34+[g]	229 325	1.382	316 927			
YP in perpetuity at 6 per cent[h]		16.667				
× PV of £1 in 33 years at 12 per cent[i]		0.024	0.400			126 771

Capital value £528 066

say £530 000

[a]Rental growth at 6.68 per cent per annum.
[b]Fixed income valued at equated yield.
[c]Deferred at equated yield.
[d]Income fixed for the next 2 years before final review.
[e]Review to agreed rental, fixed for the following 21 years.
[f]Present lease ends and the property may now be let at its rental value on the basis of 5 year rent reviews.
[g]It is acceptable to discontinue a discounted cash flow valuation at around 30 years and the rent review at the end of year 33 is a convenient point to do this.
[h]Income is now valued into perpetuity at the all risks yield from comparable information. Yield reflects rental growth of 6.68 per cent per annum and rent reviews every 5 years to produce an overall yield of 12 per cent.
[i]Deferred at equated yield. Rental growth up to this point has been shown explicitly. To defer at anything less than equated yield would incorrectly imply further rental growth.

Valuation of freehold interest using a real value approach

Rent received		£ 2 000 pa	
YP 2 years at 12 per cent [see note 1]		1.690	£ 3 380
Reversion to [see note 2]		£ 46 500 pa	
YP 21 years at 12 per cent			
[see note 3]	7.562		
x PV of £1 in 2 years at 5 per cent			
[see note 4]	0.907	6.859	£318 944
Reversion to [see note 5]		£ 37 500 pa	
YP in perpetuity at 6 per cent			
[see note 6]	16.667		
x PV of £1 in 23 years at 5 per cent			
[see note 7	0.326	5.433	£203 738
	Capital value		£526 062
	say	£530 000	

Notes

1: Rent fixed for 2 years. No growth in income, therefore valued at equated yield.

2: Reversion to agreed rental, fixed for the following 21 years.

3: Rent fixed for 21 years therefore valued at equated yield.

4: Deferred at inflation risk free yield to reflect growth in the £46 500 pa during the 2 years before review.

5: Current lease is ended and the property may now be let at its rental value on the basis of 5 year rent reviews.

6: All risks yield from comparable information. Reflects rental growth at 6.68 per cent per annum and rent reviews every 5 years to produce an overall yield of 12 per cent.

7: Deferred at inflation risk free yield to reflect growth in the £37 500 pa during the 23 years before reversion.

Leasehold interest

For the application of a real value approach, the inflation risk free yield must be re-calculated on the basis of an 18 per cent equated yield.

$$\frac{1+e}{1+g} - 1 = \frac{1.18}{1.0668} - 1 = 0.1061$$

and IRFY = 10.61 per cent.

It would be acceptable to round this off to a convenient figure in an examination.

Valuation of head leasehold interest using a discounted cash flow at equated yield approach

Year	Rent £	Amount of £1 at 6.68 per cent[b]	Inflated rent £	YP at 18 per cent[c]	PV at 18 per cent[d]	Present value £	
Capital value of rent received[a]							
1-2	28 000[e]	1	28 000	1.566	1	43 848	
3-7	37 500[f]	1.138	42 675	3.127	0.718	95 813	
8-12	42 675	1.382	58 977	3.127	0.314	57 908	
13-17	58 977	1.382	81 506	3.127	0.137	34 917	
18-22	81 506	1.382	112 641	3.127	0.060	21 134	
23	112 641	1.382	155 670	0.848	0.026	3 432	257 052[g]
less *capital value of rent paid*[a]							
1-2	2 000[h]	1	2 000	1.566	1	3 132	
3-23	46 500[i]	1.138	52 917	5.384	0.718	204 562	207 694[j]

Capital value of head leasehold interest   £49 358[k]

say £50 000

[a]Rent received and rent paid valued separately so that the different growth prospects of the two incomes may be taken into account.
[b]Rental growth at 6.68 per cent per annum.
[c]Fixed income valued at equated yield.
[d]Deferred at equated yield.
[e]Rent fixed for next 2 years until review.
[f]Review to current rental value on basis of 5 year reviews.
[g]Capital value of rent received by head leaseholder.
[h]Rent fixed for next 2 years of head lease.
[i]Review to agreed rent, fixed for last 21 years of the head lease.
[j]Capital value of rent paid by head leaseholder.
[k]Capital value of rent paid by head leaseholder is deducted from capital value of rent received, to give capital value of head leasehold interest.

Capital value of head leasehold interest using a real value approach

Capital value of rent received:
[see note 1]

Rent received	£28 000 pa	
YP 2 years at 18 per cent [see note 2]	1.566	£43 848
Reversion to [see note 3]	£37 500 pa	

$$\text{YP 21 years at 10.61 per cent} \times \frac{\text{YP 5 years at 18 per cent}}{\text{YP 5 years at 10.61 per cent}}$$
[see note 4]

$$= 8.291 \times \frac{3.127}{3.732} \quad = 6.947$$

× PV of £1 in 2 years at 10.61 per cent [see note 5]	0.817	5.676	£212 850	£256 698 [see note 6]

less *capital value of rent paid:*
[see note 1]

Rent paid		£ 2 000 pa		
YP 2 years at 18 per cent [see note 7]		1.566	£ 3 132	
Reversion to [see note 8]		£46 500 pa		
YP 21 years at 18 per cent [see note 9]	5.384			
× PV of £1 in 2 years at 10.61 per cent [see note 10]	0.817	4.399	£204 554	£207 686 [see note 11]

Capital value of head leasehold interest [see note 12]	£ 49 012

say £50 000

Notes

1: See note[a], head leasehold interest, discounted cash flow approach.

2: Rent fixed for 2 years. No growth in income, therefore valued at equated yield.

3: Review to current rental value on basis of 5 year rent reviews.

4: Three YPs reflect total length of unexpired head lease, rental growth of 6.68 per cent per annum and rent reviews every 5 years, to give an overall yield of 18 per cent.

5: Deferred at inflation risk free yield to reflect growth in the £37 500 pa during the first 2 years.

6: Capital value of rent received by head leaseholder.

7: Rent fixed for 2 years, therefore valued at equated yield.

8: Review to agreed rent for remaining 21 years of head lease.

9: Rent fixed for 21 years, therefore valued at equated yield.

10: Deferred at inflation risk free yield to reflect growth in the £46 500 pa during the first 2 years.

11: Capital value of rent paid by head leaseholder.

12: Capital value of rent paid by head leaseholder is deducted from capital value of rent received to give capital value of head leasehold interest.

Question 9.5

Before the required valuations may be carried out, information relating to the comparable properties must be analysed.

This analysis should reveal the all risks yield, the Zone A rental value per m^2 of the shops, the rate of rental growth per annum, the inflation risk free yield and the equivalent yield.

A further unknown is the equated yield, necessary for the valuation using real value approach, and this may be estimated using the information given regarding the yield from gilts. The question states that long dated gilts are yielding 10 per cent and, using the accepted method of adding 2 per cent to this yield, gives an equated yield of 12 per cent (see Question 9.2).

Analysis of information from Property 1

To find Zone A rental value per m^2, using two 5 m zones and a remainder and halving back.

Let Zone A rental value per m^2 = £X.

Zone A   = 7 m × 5 m × $X = 35X$
Zone B   = 7 m × 5 m × $\frac{1}{2}X = 17.5X$
Remainder = 7 m × 6.5 m × $\frac{1}{4}X = 11.375X$
$$\text{Rental value } = 63.875X$$

$$63.875X = £31\ 500 \text{ pa}$$
$$\text{and} \quad X = £493.15$$
$$\text{say} \quad = £493 \text{ per } m^2$$

The information from this property may also be used to discover the all risks yield, on the basis of 5 year reviews, since not only is it a recent letting, but the freehold interest has also just been sold.

$$\text{All risks yield} = \frac{£31\ 500}{£630\ 000} \times 100 = 5 \text{ per cent}$$

The all risks yield may now be analysed, to ascertain the rate of rental growth per annum implied by the acceptance of a yield of 5 per cent, using the formula:

$$(1+g)^n = \frac{\text{YP in perpetuity at } i - \text{YP } n \text{ years at } e}{\text{YP in perpetuity at } i \times \text{PV of £1 in } n \text{ years at } e}$$
(see Question **9.1**)

$$(1+g)^5 = \frac{\text{YP in perpetuity at 5 per cent} - \text{YP 5 years at 12 per cent}}{\text{YP in perpetuity at 5 per cent} \times \text{PV of £1 in 5 years at 12 per cent}}$$

$$(1+g)^5 = \frac{20 - 3.605}{20 \times 0.567} = \frac{16.395}{11.340} = 1.446$$

$$\sqrt[5]{1.446} = 1 + g = 1.0765$$

$$g = 0.0765$$

The implied rate of rental growth is therefore 7.65 per cent per annum.

This information, together with the equated yield will reveal the inflation risk free yield (IRFY), using the formula:

$$\frac{1+e}{1+g} - 1 = i \text{ (see Question 9.1)}$$

$$\frac{1.12}{1.0765} - 1 = 0.0404$$

and IRFY = 4.04 per cent.

It would be acceptable, in an examination, to round this off to 4 per cent.

Analysis of information from Property 2

The Zone A rental value from Property 1 is first applied, in order to find the current rental value of Property 2. Once again, two 5 m zones and a remainder are used, so that both properties are considered on the same basis.

Zone A   = 6 m x 5 m x £493   = £14 790
Zone B   = 6 m x 5 m x £246.5   = £ 7 395
Remainder = 6 m x 6 m x £123.25 = £ 4 437

Current rental value on 5 year reviews   £26 622

              say   £26 625 pa

This figure, together with the current rent passing and the recent sale price of the freehold interest, will enable derivation of the equivalent yield.

This may be calculated by the formula:

$$\frac{\text{Present income} + \text{annual equivalent of gain on reversion}}{\text{Price}} \times 100$$

However, this can be a time-consuming, lengthy calculation. The equivalent yield is the overall yield from an investment in current rental terms. Therefore, in an examination situation, the equivalent yield is more easily discovered using discounted cash flow analysis, calculating the internal rate of return of the current income flow.

A test yield must be selected and, in this case, 4.5 per cent has been chosen.

Rent received	£15 750 pa	
YP 4 years at 4.5 per cent	3.587	£ 56 495
Reversion to	£26 625 pa	
YP in perpetuity deferred 4 years at 4.5 per cent	18.635	£496 157
		£552 652
less Purchase price		£495 000
Net present value at 4.5 per cent		+ £ 57 652

The internal rate of return (equivalent yield) is therefore greater than 4.5 per cent and a yield of 5 per cent will be tested.

Rent received	£15 750 pa	
YP 4 years at 5 per cent	3.546	£ 55 850
Reversion to	£26 625 pa	
YP in perpetuity deferred 4 years at 5 per cent	16.454	£438 088
		£493 938
less Purchase price		£495 000
Net present value at 5 per cent		− £ 1 062

The equivalent yield must lie between 4.5 and 5 per cent and may be discovered using similar triangles.

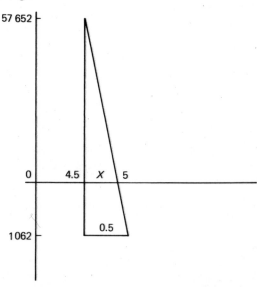

$$\frac{X}{57\,652} = \frac{0.5}{58\,714}$$

$$58\,714X = 28\,826$$

$$X = 0.49$$

Equivalent yield = 4.5 per cent + 0.49 per cent

= 4.99 per cent.

The final piece of information necessary to carry out the required valuations is the current rental value of the subject property. This is derived using the Zone A rental value from Property 1, again adopting two 5 m zones and a remainder.

```
Zone A      = 6 m ×   5 m × £493      = £14 790
Zone B      = 6 m ×   5 m × £246.5    = £  7 395
Remainder   = 6 m × 7.5 m × £123.25   = £  5 546
                                        £27 731
```

Current rental value of subject property on basis of 5 year reviews, say £27 750 pa.

Valuation of freehold interest using a traditional approach

Rent received	£23 500 pa	
YP 2 years at 4 per cent [see note 1]	1.886	£ 44 321
Reversion to [see note 2]	£27 750 pa [see note 3]	
YP in perpetuity deferred 2 years at 5 per cent [see note 4]	18.141	£503 413
Capital value		£547 734
say	£550 000	

Notes

1: Traditional yield pattern. Term income is below current rental value, therefore considered to be more secure and valued at 1 per cent below all risks yield.
2: Next review is in 2 years' time.
3: Current rental value on 5 year review pattern.
4: All risks yield on basis of 5 year reviews, from comparable information.

The anomaly that becomes apparent in this valuation is that the current rental value and all risks yield obtained by analysis of transactions from comparable properties are both on the basis of a 5 year rent review pattern, whereas the subject property is let on a 3 year review pattern. There might therefore be a temptation to make some adjustment to the yield to allow for this fact, since the receipt of an income reviewed every 3 years is more attractive than an income reviewed every 5 years.

However, the valuation should not require any manipulation since the current rental value and the all risks yield are both on the same 5 year review basis. The capital value based on 5 year reviews should equal the capital value on 3 year reviews, as long as the basis of the all risks yield matches that of the rental value.

If the yield is manipulated on to 3 year review terms, the current rental value must also be adjusted on to this basis and the traditional approach has no device for carrying out such an adjustment.

Any such manipulation must inevitably be intuitive and subjective, unless the valuer has knowledge of comparable transactions for guidance.

A further problem is the use of 4 per cent to value the term income. The low yield implies growth, whereas the £23 500 pa is fixed. The effect is not drastic in this instance, since the income is fixed for only 2 years, but if there were a long unexpired term, it could result in a significant over-valuation.

Some under-valuation of the reversion also occurs, since this value is deferred at the all risks yield, which does not allow sufficiently for growth in the current rental value during the period before reversion. Again, this problem increases with the length of the unexpired term.

For further consideration of the traditional method of valuation, see Question 9.2.

Valuation of freehold interest using equivalent yield approach

Rent received		£23 500 pa	
YP 2 years at 4.99 per cent [see note 1]		1.86	£ 43 710
Reversion to [see note 2]		£27 750 pa	
		[see note 3]	
YP in perpetuity at 4.99 per cent [see note 1]	20.040		
× PV of £1 in 2 years at 4.99 per cent [see note 1]	0.907	18.176	£504 384
	Capital value		£548 094

say £550 000

Notes

1: Valuation of both term and reversion and deferment of the reversion all carried out at the same yield — the equivalent yield.

2: Next review in 2 years' time.

3: Current rental value on 5 year review pattern.

Once again, there could be problems in applying the information analysed to the valuation of the freehold interest in the subject property. The situation involved in the latter does not coincide with that of the comparable. The equivalent yield of 4.99 per cent was derived from the lease of a property having an unexpired term of 4 years (compared with 2 years in the case of the subject property) and a current rental value on the basis of a 5 year review pattern (subject property let on 3 year reviews). Should, therefore, the equivalent yield be manipulated to allow for the differing circumstances of comparable and subject properties? If so, by how much should the yield be changed and in what direction? Unless ideal comparables are available, manipulation must be subjective and can result in wide variations in capital value.

The use of the same, low yield throughout the valuation also introduces the problem of growth implications, as explained above in respect of the traditional approach.

Valuation of freehold interest using a real value approach

In this instance, the valuer need not consider any intuitive adjustments, since the individual circumstances involved in the letting of the subject property, may be incorporated in the valuation.

Firstly, the current rental value may be adjusted on to the basis of a 3 year review pattern, using the formula:

$$X \times \frac{\text{YP } n \text{ years at } e}{\text{YP } n \text{ years at } i} \times \frac{\text{YP } N \text{ years at } l}{\text{YP } N \text{ years at } e} = y$$
(see Question 9.1)

Thus, rental value on a 3 year review pattern

$$= £27\,750 \times \frac{\text{YP 5 years at 12 per cent}}{\text{YP 5 years at 4.04 per cent}} \times \frac{\text{YP 3 years at 4.04 per cent}}{\text{YP 3 years at 12 per cent}}$$

$$= £27\,750 \times \frac{3.605}{4.447} \times \frac{2.773}{2.402} = £25\,970 \text{ pa}$$

The valuation then appears as:

Rent received	£23 500 pa	
YP 2 years at 12 per cent [see note 1]	1.690	£39 715
Reversion to [see note 2]	£25 970 pa	

$$\text{YP in perpetuity at 4.04 per cent} \times \frac{\text{YP 3 years at 12 per cent}}{\text{YP 3 years at 4.04 per cent}}$$
[see note 3]

$$= 24.752 \times \frac{2.402}{2.773} = 21.440 \text{ [see note 4]}$$

× PV of £1 in 2 years at 4.04 per cent [see note 5]	0.924	19.811	£514 492
	Capital value		£554 207

say £550 000

Notes

1: Rent fixed for 2 years. No growth in income, therefore valued at equated yield.

2: Review to current rental value on 3 year reviews.

3: Three YPs reflect rental growth of 7.65 per cent per annum and a 3 year review pattern, to produce an overall yield of 12 per cent.

4: YP in perpetuity produced by 3 YPs. Represents an all risks yield on a 3 year review pattern of 4.66 per cent.

5: Deferred at inflation risk free yield to reflect growth in the £25 970 pa during the 2 years before review.

Alternatively, the reversion may be valued on the basis of a 5 year review pattern, although the valuation above reflects the true situation of the letting.

The capital value of the rent on reversion should be the same whether it is valued assuming 3 year or 5 year reviews, as long as the yield and the rental value are adjusted accordingly. This is the theory relied upon by the constant rent formula. Allowance for the review pattern is made in the yield applied in the valuation. In the case of rental value on 3 year reviews, the YP in perpetuity of 21.44 represents a yield of 4.66 per cent, whereas when the current rental value is expressed in terms of 5 year reviews, the yield used is 5 per cent. Using the latter, the valuation becomes:

Rent received		£23 500 pa	
YP 2 years at 12 per cent		1.690	£ 39 715
Reversion to		£27 750 pa	
YP in perpetuity at 5 per cent	20		
× PV of £1 in 2 years at 4.04 per cent	0.924	18.48	£512 820
	Capital value		£552 535

say £550 000

This confirms the point made earlier regarding the traditional approach, where it was argued that the all risks yield on reversion should not be manipulated, unless the current rental value is also adjusted on to the appropriate review pattern.

In conclusion, the student should briefly comment on the overall approaches of the three valuation methods.

Term and reversion layout is adopted in each case.

Traditional and equivalent yield methods use an implicit approach to reflect rental growth in the yields, sometimes incorrectly. In the equivalent yield method, the same yield is used throughout the valuation, whereas the traditional approach is to value the reversion at the all risks yield, with the term yield adjusted in a downwards direction to reflect the assumed security of income. Objectivity in analysis is common to both methods, although subjectivity may be introduced at the valuation stage should valuer intuition be necessary to reflect differences between comparables and subject property.

The real value method is explicit regarding income growth, or lack of it. There is a subjective choice of equated yield in the initial stages of analysis, but the valuation is entirely objective, requiring no intuitive manipulation by the valuer. Crosby has shown that the subjective choice of equated yield is not crucial, particularly if the unexpired term is not in excess of 12 years. (*Journal of Valuation, Vol. 4, 1986. 'The Application of Equated Yield and Real Value Approaches to Market Valuation: 2. Equivalent Yield or Equated Yield Approaches?'*.)

Traditional and equivalent yield methods both use growth-implicit yields in a situation where there is no growth, that is, to value the term income. In the example under consideration, the effect is not too great, since the term is of only 2 years' duration, but the longer the unexpired term, the greater is the over-valuation. Both of these methods virtually ignore growth in the current rental value during the years before reversion, by deferring reversionary value at the same yield as that used to value. This under values the reversion and, again, the effect increases with the length of unexpired term.

To value the term income, a yield reflecting a fixed income is used in the real value approach, derived by increasing the yield from long dated gilts by approximately 2 per cent. Thus a fixed income from property is valued by comparison with other fixed income investments. Reversion is valued using the all risks yield, as in the traditional approach, but the 3 YPs can be used to deal with differing rent review patterns. Growth in the current rental value during the period before reversion is also accounted for, either by reflecting this in the yield used to defer the reversionary value (the inflation risk free yield), or by reflecting growth explicitly in the rental value and deferring at the equated yield.

Finally, it is of interest to note that, despite the different approaches, the 3 valuations produce roughly the same answer in this instance. The valuer must ultimately decide which method has the most logically sound base upon which the valuation may be defended.

Question 9.6

Freehold interest

Valuation of freehold interest using a real value approach
In order to carry out this valuation, it is necessary to determine the rate of rental growth implied by the acceptance of a rack rented capitalisation rate of 5 per cent, the inflation risk free yield and the equated yield.

The equated yield will be determined by adopting the accepted practice of increasing the yield from long dated gilt edged securities by 2 per cent. The question states that long dated gilts are yielding 10 per cent, therefore an equated yield of 12 per cent will be assumed (see Question 9.2).

Analysis to discover the implied growth rate now follows, using the formula:

$$(1+g)^n = \frac{\text{YP in perpetuity at } i - \text{YP } n \text{ years at } e}{\text{YP in perpetuity at } i \times \text{PV of £1 in } n \text{ years at } e}$$

(see Question 9.1)

$$(1+g)^5 = \frac{\text{YP in perpetuity at 5 per cent} - \text{YP 5 years at 12 per cent}}{\text{YP in perpetuity at 5 per cent} \times \text{PV of £1 in 5 years at 12 per cent}}$$

$$(1+g)^5 = \frac{20 - 3.605}{20 \times 0.567} = \frac{16.395}{11.340} = 1.446$$

$$\sqrt[5]{1.446} = 1 + g = 1.0765$$

$$g = 0.0765$$

The implied rate of rental growth is therefore 7.65 per cent per annum.

The inflation risk free yield (IRFY) is now calculated, using the formula:

$$\frac{1+e}{1+g} - 1 = i \quad \text{(see Question 9.1)}$$

$$\frac{1.12}{1.0765} - 1 = 0.0404$$

and IRFY = 4.04 per cent.

It would be acceptable in an examination, to round this off to 4 per cent.

Valuation

Rent received	£14 000 pa	
YP 18 years at 12 per cent [see note 1]	7.25	£101 500
Reversion to [see note 2]	£65 000 pa	
YP in perpetuity at 5 per cent [see note 3]	20	
x PV of £1 in 18 years at 4.04 per cent [see note 4]	0.490 9.80	£637 000
Capital value		£738 500

say £740 000

Notes

1: Rent received is fixed for 18 years. No growth in income, therefore valued at equated yield.

2: Reversion to current rental value on 5 year reviews.

3: Rack rented capitalisation rate from comparable transactions. Reflects rental growth at 7.65 per cent per annum and 5 year rent reviews, to produce an overall yield of 12 per cent.

4: Deferred at inflation risk free yield to reflect growth in rental value during the 18 years before reversion.

Valuation of freehold interest using the rational approach

The variables which need to be determined before proceeding with this valuation are the risk adjusted opportunity cost of capital and the annual rate of rental growth implied by the acceptance of a 5 per cent rack rented capitalisation rate.

Estimation of the risk adjusted opportunity cost of capital results from the same reasoning as that applied in determining the equated yield for the real value approach. A 12 per cent yield will therefore be adopted in this instance.

The formula used in the rational model to calculate the implied growth rate, differs from that used in the real value model, but the result is the same.

$$g = \left[\left(\frac{d-y}{d} \right) \left((1+d)^t - 1 \right) + 1 \right]^{\frac{1}{5}} - 1$$

where
  g = annual growth rate
  d = risk adjusted opportunity cost of capital
  y = rack rented capitalisation rate
  t = number of years in rent review pattern
$\left. \right\}$ all expressed as decimals

$$g = \left[\left(\frac{0.12 - 0.05}{0.12} \right) \left((1.12)^5 - 1 \right) + 1 \right]^{\frac{1}{5}} - 1$$

$$\text{or } (1+g)^5 = \left[\left(\frac{0.12 - 0.05}{0.12} \right) \left((1.12)^5 - 1 \right) \right] + 1$$

$$(1+g)^5 = (0.5833)(0.7623) + 1 = 1.445$$

$$\sqrt[5]{1.445} = 1 + g = 1.0765$$

$$g = 0.0765$$

The implied rate of rental growth is therefore 7.65 per cent per annum.

A term and reversion freehold valuation by the rational approach takes the following form:

$$C \doteq \left(\frac{r}{d} - \frac{r}{d(1+d)^n} \right) + \frac{R(1+g)^n}{y(1+d)^n}$$

where
  C = capital value
  r = current rental income
  R = estimated current rental value
  y = rack rented capitalisation rate
  n = number of years to next review
  d = risk adjusted opportunity cost of capital.

Put into a more familiar format, the valuation appears as:

Rent received		£ 14 000 pa		
YP in perpetuity at 12 per cent				
[see note 1		8.333	£116 662	
Less Rent received		£ 14 000 pa		
YP in perpetuity at 12 per cent				
[see note 2]	8.333			
x PV of £1 in 18 years at				
12 per cent	0.130	1.083	£ 15 162	£101 500
Reversion to [see note 3]		£ 65 000 pa		
x Amount of £1 in 18 years at				
7.65 per cent [see note 4]		3.769		
Estimated rental value in				
18 years [see note 5]		£244 985		
YP in perpetuity at 5 per cent				
[see note 6]	20			
x PV of £1 in 18 years at				
12 per cent [see note 7]	0.130	2.6		£636 961
		Capital value		£738 461

say £740 000

Notes

1: Current income is first valued in perpetuity. Valuation at risk adjusted opportunity cost of capital reflects fixed income.

2: Current income is valued in perpetuity, deferred for the 18 years of the unexpired term. Resultant figure is deducted from capital value of the income in perpetuity. The effect is the same as simply valuing £14 000 pa for 18 years at the risk adjusted opportunity cost of capital.

3: Reversion to current rental value on basis of 5 year rent reviews.

4: Rental value escalated at implied rate of rental growth.

5: Estimated rental value in 18 years' time, if rents grow at 7.65 per cent per annum. Rental growth is shown explicitly in the valuation.

6: Valuation into perpetuity at rack rented capitalisation rate of 5 per cent, from comparable information.

7: Deferred at risk adjusted opportunity cost of capital, since rental growth has been made explicit. To defer at a lower yield would incorrectly imply further growth in rental value.

Head leasehold interest

Valuation of head leasehold interest using a real value approach

An equatcd yield of 12 per cent was used in the valuation of the freehold interest.

To allow for the extra risks that are perceived in leasehold interests, a higher equated yield is necessary and, in this instance, 18 per cent will be adopted. This means that the inflation risk free yield must be recalculated.

$$\frac{1+e}{1+g} - 1 = \frac{1.18}{1.0765} - 1 = 0.0961$$

and IRFY = 9.61 per cent.

In an examination, it would be acceptable to round this off to a more convenient figure.

In the valuation, to account for the differing growth potentials of the head lease and sub-lease rents, they are capitalised separately. The method entirely abandons the traditional concept of valuing a profit rent.

Capital value of rent received

Rent received [see note 1]	£45 000 pa	
YP 3 years at 18 per cent [see note 2]	2.174	£ 97 830
Reversion to [see note 3]	£65 000 pa	

$$\text{YP 15 years at 9.61 per cent} \times \frac{\text{YP 5 years at 18 per cent}}{\text{YP 5 years at 9.61 per cent}}$$

[see note 4]

$$= 7.778 \times \frac{3.127}{3.829} = 6.352$$

× PV of £1 in 3 years at 9.61 per cent [see note 5]	0.759	4.821	£313 365
[see note 6]			£411 195

less *capital value of rent paid:*

Rent paid	£14 000 pa	
YP 18 years at 18 per cent [see note 7]	5.273	£ 73 822
		[see note 8]
Capital value [see note 9]		£337 373

say £340 000

Notes

1: Rent under sublease fixed until next review in 3 years.

2: Fixed rent. No growth in income, therefore valued at equated yield.

3: Reversion to current rental value on basis of 5 year reviews.

4: Three YPs reflect length of unexpired term of lease, rental growth of 7.65 per cent per annum, with rent reviews every 5 years to produce an overall yield of 18 per cent.

5: Deferred at inflation risk free yield to reflect growth in rental value during the 3 years before reversion.

6: Capital value of rent received by head leaseholder.

7: Rent under head lease fixed for 18 years, therefore valued at equated yield.

8: Capital value of rent paid by head leaseholder.

9: Capital value of rent paid by head leaseholder is deducted from capital value of rent received, to give capital value of head leasehold interest.

Valuation of head leasehold interest using rational approach

The valuation of the head leasehold interest using the rational model takes the following form:

$$C = \left(\frac{r}{d} - \frac{r}{d(1+d)^n} \right) + \frac{R(1+g)^n}{y(1+d)^n} - \frac{R(1+g)^N}{y(1+d)^N}$$

where

C = capital value

n = number of years to next review

d = risk adjusted opportunity cost of capital

y = rack rented capitalisation rate

g = implied rate of annual rental growth

r = net initial rent passing

R = estimated net rack rent

N = length of leasehold interest.

The risk adjusted opportunity cost of capital is assumed to be the same yield as that adopted for the equated yield in the real value approach to the head leasehold interest, that is, 18 per cent.

In a more familiar format, the valuation appears as:

Rent received	£45 000 pa	
less Rent paid	£14 000 pa	
Profit rent [see note 1]	£31 000 pa	
YP in perpetuity at 18 per cent [see note 2]	5.555 £172 205	
Less Rent received	£45 000 pa	
less Rent paid	£14 000 pa	
Profit rent	£31 000 pa	
YP in perpetuity at 18 per cent 5.555		
x PV of £1 in 3 years at 18 per cent 0.609	3.383 £104 873	£67 332
[see note 3]		

Plus Rent received [see note 4]		£65 000 pa	
less Rent paid		£14 000 pa	
Profit rent		£51 000 pa	
x Amount of £1 in 3 years at 7.65 per cent [see note 5]		1.2475	
Estimated profit rent in 3 years [see note 6]		£636 22 pa	
YP in perpetuity at 5 per cent [see note 7]	20		
x PV of £1 in 3 years at 18 per cent [see note 8]	0.609	12.18 £774 916	
			[see note 9]
Less Rent received		£ 65 000 pa	
less Rent paid		£ 14 000 pa	
Profit rent		£ 51 000 pa	
x Amount of £1 in 18 years at 7.65 per cent [see note 10]		3.769	
Estimated profit rent in 18 years [see note 11]		£192 219 pa	
YP in perpetuity at 5 per cent [see note 12]	20		
x PV of £1 in 18 years at 18 per cent [see note 13]	0.051	1.02 £196 063 £578 853	
		[see note 14]	
	Capital value		£646 185

say £645 000

Notes

1: Method uses conventional profit rent or net income as basis of valuation.

2: Profit rent valued in perpetuity at risk adjusted opportunity cost of capital.

3: Profit rent is again valued in perpetuity, but deferred for the length of the term (3 years). Resultant figure is deducted from capital value in perpetuity, to give capital value of profit rent for first 3 years. Put more simply: value in perpetuity minus value of reversion equals value of term. The effect is the same as valuing the fixed profit rent for 3 years at the risk adjusted opportunity cost of capital:

Profit rent	£31 000 pa
YP 3 years at 18 per cent	2.174
Capital value	£67 394 (error due to rounding)

4: Reversion to current rental value on basis of 5 year reviews.

5: Rental growth over 3 years at 7.65 per cent per annum.

6: Estimated profit rent in 3 years. Profit rent is assumed to grow at 7.65 per cent per annum.

7: Yield derived from transactions involving comparable freehold properties.

8: Deferred at risk adjusted opportunity cost of capital. Growth has been shown explicitly in the income, and to defer at a lower yield would incorrectly imply further rental growth.

9: Estimated profit rent in 3 years' time is valued in perpetuity.

10: Rental growth over 18 years at 7.65 per cent per annum.

11: Estimated profit rent in 18 years' time assuming growth at 7.65 per cent per annum.

12: See note 7.

13: See note 8.

14: Capital value of reversion in 18 years' time is deducted from capital value in perpetuity in 3 years' time, to produce capital value of the last 15 years of the head leasehold interest.

An anomaly is immediately obvious. The real value model values the head leasehold interest at approximately £340 000, whereas the rational valuation is nearer £645 000 — a difference of over £300 000.

One problem with the rational approach is its reliance on the capitalisation rate, with no device for amending it. It is normally perceived that some adjustment is necessary for the extra risks attached to investment in leasehold interests. This is the reason for adopting 18 per cent as equated yield and as the risk adjusted opportunity cost of capital.

However, in the rational model, this results in a mixture of interest rates within the valuation, when the 18 per cent is used together with the 5 per cent rack rented capitalisation rate. A yield of 5 per cent reflects rental growth of 7.65 per cent per annum, but, from the freehold analysis and valuation, it is known that 5 per cent also implies a discount rate of 12 per cent. Thus, the valuation has been carried out at 18 per cent, but the valuation in perpetuity is calculated at 12 per cent.

To correct this, it would be necessary to adjust the capitalisation rate so that it represents a risk adjusted opportunity cost of capital of 18 per cent and a growth rate of 7.65 per cent per annum. Unfortunately, the rational model has no device for making such an adjustment.

In presenting the model, some attempt to overcome this difficulty was made, in effect by adjusting the growth rate downwards when valuing a leasehold interest. This is achieved by adjusting the capitalisation rate in the same way as would be done in a conventional valuation, that is, by the addition of, say, 1 per cent to the freehold capitalisation rate, leaving the risk adjusted opportunity cost of capital unchanged from that used in the freehold valuation. The implied growth rate is then re-calculated on this basis.

$$g = \left[\left(\frac{0.12 - 0.06}{0.12} \right) \left((1.12)^5 - 1 \right) + 1 \right]^{\frac{1}{5}} - 1$$

or

$$(1 + g)^5 = \left[\left(\frac{0.12 - 0.06}{0.12}\right)\left((1.12)^5 - 1\right) + 1\right]$$

$$= (0.5)(0.7623) + 1 = 1.381$$

$$\sqrt[5]{1.381} = 1 + g = 1.0667$$

$$g = 0.0667$$

The implied rate of rental growth is therefore 6.67 per cent per annum.

A consideration of the validity of changing the rate of growth will be returned to later.

The valuation now becomes:

Profit rent		£31 000 pa		
YP in perpetuity at 12 per cent		8.333	£258 323	
Less Profit rent		£31 000 pa		
YP in perpetuity at 12 per cent	8.333			
x PV of £1 in 3 years at				
12 per cent	0.712	5.933	£183 923	£74 400
Plus Profit rent		£51 000 pa		
x Amount of £1 in 3 years at				
6.67 per cent		1.214		
Estimated profit rent in 3 years		£61 914 pa		
YP in perpetuity at 6 per cent	16.667			
x PV of £1 in 3 years at				
12 per cent	0.712	11.867	£734 733	
Less Profit rent		£51 000 pa		
x Amount of £1 in 18 years at				
6.67 per cent		3.197		
Estimated profit rent in 18 years		£163 047 pa		
YP in perpetuity at 6 per cent	16.667			
x PV of £1 in 18 years at				
12 per cent	0.130	2.167	£353 323	£381 410
		Capital value		£455 810

say £455 000

There is still approximately £120 000 difference between this valuation and that produced by the real value approach.

A further difficulty involved in this model is that no allowance is made for the gearing of the two leases. It is claimed that the method is explicit regarding growth, but it is assumed in the model that profit rent is growing at the same rate as rental value, which it patently is not. The incomes from the two leases have completely different growth potentials — the head lease is at a fixed rent, whereas the sub-lease is on a rising rent A possible solution might be to adopt the format of the real value approach capitalising the head lease and sub-lease rents completely separately.

This would appear as follows:

Capital value of rent received:

Rent received		£45 000 pa		
YP in perpetuity at 12 per cent		8.333	£374 985	
Less Rent received		£45 000 pa		
YP in perpetuity at 12 per cent	8.333			
x PV of £1 in 3 years at				
12 per cent	0.712	5.933	£266 985	£108 000
Plus Rent received		£65 000 pa		
x Amount of £1 in 3 years at				
6.67 per cent		1.214		
Estimated rental value in 3 years		£78 910 pa		
YP in perpetuity at 6 per cent	16.667			
x PV of £1 in 3 years at				
12 per cent	0.712	11.867	£936 425	
Less Rent received		£65 000 pa		
x Amount of £1 in 18 years at				
6.67 per cent		3.197		
Estimated rental value in 18 years		£207 805 pa		
YP in perpetuity at 6 per cent	16.667			
x PV of £1 in 18 years at				
12 per cent	0.130	2.167	£450 313	£486 112
Capital value of rent received				£594 112

less *capital value of rent paid:*

rent paid		£ 14 000 pa		
YP in perpetuity at 12 per cent		8.333	£116 662	
Less Rent paid		£ 14 000 pa		
YP in perpetuity at 12 per cent	8.333			
x PV of £1 in 18 years at				
12 per cent	0.130	1.083	£ 15 162	£101 500
Capital value				£492 612

say £490 000

This approach does not appear to help the situation. There still seems to be some remaining problem in the yields employed.

It may help to return to what might be the crux of this difficulty — the alteration of the capitalisation rate. Increasing this rate from 5 per cent to 6 per cent had the effect of reducing the implied growth rate by almost 1 per cent to 6.67 per cent. But is it correct to amend the growth rate? Surely, if the rental value is increasing, it will increase at the same rate whether it is being received by the freeholder or the leaseholder.

Performing a rational valuation of the head leasehold interest using a discount rate of 12 per cent, a rack rented capitalisation rate of 5 per cent and a growth rate of 7.65 per cent per annum, produces the following result:

Profit rent		£ 31 000 pa		
YP in perpetuity at 12 per cent		8.333	£258 323	
Less Profit rent		£ 31 000 pa		
YP in perpetuity at 12 per cent	8.333			
× PV of £1 in 3 years at				
12 per cent	0.712	5.933	£183 923	£ 74 400
Plus Profit rent		£ 51 000 pa		
× Amount of £1 in 3 years at				
7.65 per cent		1.2475		
Estimated profit rent in 3 years		£ 63 622 pa		
YP in perpetuity at 5 per cent	20			
× PV of £1 in 3 years at				
12 per cent	0.712	14.24	£905 977	
Less Profit rent		£ 51 000 pa		
× Amount of £1 in 18 years at				
7.65 per cent		3.769		
Estimated profit rent in 18 years		£192 219 pa		
YP in perpetuity at 5 per cent	20			
× PV of £1 in 18 years at				
12 per cent	0.130	2.6	£499 769	£406 208
		Capital value		£480 608

say £480 000

However, this valuation still maintains the assumption that the profit rent is growing at the same rate as current rental value. This does not reflect the true circumstances and, valuing the head lease and sub-lease rents separately, but adopting the same yields as in the above valuation, produces the following result.

Capital value of rent received:

Rent received	£ 45 000 pa		
YP in perpetuity at 12 per cent	8.333	£374 985	

Less Rent received		£ 45 000 pa		
YP in perpetuity at 12 per cent	8.333			
x PV of £1 in 3 years at				
12 per cent	0.712	5.933	£266 985	£108 000

Plus Rent received	£ 65 000 pa	
x Amount of £1 in 3 years at		
7.65 per cent	1.2475	

Estimated rental value in 3 years	£ 81 087 pa		
YP in perpetuity at 5 per cent	20		
x PV of £1 in 3 years at			
12 per cent	0.712	14.24	£1 154 679

Less Rent received	£ 65 000 pa	
x Amount of £1 in 18 years at		
7.65 per cent	3.769	

Estimated rental value in 18 years		£244 985 pa		
YP in perpetuity at 5 per cent	20			
x PV of £1 in 18 years at				
12 per cent	0.130	2.6	£636 961	£517 718
				£625 718

less *capital value of rent paid:*

Rent paid	£ 14 000 pa		
YP in perpetuity at 12 per cent	8.333	£116 662	

Less Rent paid		£ 14 000 pa		
YP in perpetuity at 12 per cent	8.333			
x PV of £1 in 18 years at				
12 per cent	0.130	1.083	£ 15 162	£101 500
	Capital value			£524 218

say £525 000

An alternative would be to carry out the valuation, again showing growth in rental value rather than profit rent, but adopting the rational model format. This might be more desirable since it retains the identity of the rational approach, and would appear as follows:

Rent received		£ 45 000 pa			
less Rent paid		£ 14 000 pa			
Profit rent		£ 31 000 pa			
YP in perpetuity at 12 per cent		8.333	£258 323		
Less Rent received		£ 45 000 pa			
less Rent paid		£ 14 000 pa			
Profit rent		£ 31 000 pa			
YP in perpetuity at 12 per cent	8.333				
x PV of £1 in 3 years at					
12 per cent	0.712	5.933	£183 923	£ 74 400	
Plus Rent received		£ 65 000 pa			
x Amount of £1 in 3 years at					
7.65 per cent		1.2475			
Estimated rental value in 3 years		£ 81 087 pa			
YP in perpetuity at 5 per cent	20				
x PV of £1 in 3 years at					
12 per cent	0.712	14.24	£1 154 679		
Less Rent paid		£ 14 000 pa			
YP in perpetuity at 12 per cent	8.333				
x PV of £1 in 3 years at					
12 per cent	0.712	5.933	£ 83 062	£1 071 617	
Less Rent received		£ 65 000 pa			
x Amount of £1 in 18 years at					
7.65 per cent		3.769			
Estimated rental value in 18 years		£244 985 pa			
YP in perpetuity at 5 per cent	20				
x PV of £1 in 18 years at					
12 per cent	0.130	2.6	£636 961		
Less Rent paid		£ 14 000 pa			
YP in perpetuity at 12 per cent	8.333				
x PV in £1 in 18 years at					
12 per cent	0.130	1.083	£ 15 162	£621 799	£449 818
	Capital value				£524 218

say £525 000

In essence this is the same as valuing rent received and rent paid completely separately, but follows more closely the rational model format.

As a comparison, it is interesting to note the answer produced by the real value approach if an equated yield of 12 per cent is adopted to value the head leasehold interest.

Capital value of rent received:

Rent received	£45 000 pa	
YP 3 years at 12 per cent	2.402	£108 090
Rent received	£65 000 pa	

$$\text{YP 15 years at 4.04 per cent} \times \frac{\text{YP 5 years at 12 per cent}}{\text{YP 5 years at 4.04 per cent}}$$

$$= 11.087 \times \frac{3.605}{4.447} = 8.988$$

x PV of £1 in 3 years at 4.04 per cent	0.888	7.981	£518 765
			£626 855

less *capital value of rent paid:*

Rent paid	£14 000 pa	
YP 18 years at 12 per cent	7.250	£101 500
Capital value		£525 355

say £525 000

At last, some agreement appears in the valuation of the head leasehold interest by the two different approaches. However, assuming that some risk adjustment to the yield is required in valuing a leasehold interest, the 12 per cent employed above may be too low.

If the required yield is taken to be 18 per cent, as used in the real value approach, in order to apply the rational model, it is necessary to determine the rack rented capitalisation rate that will produce an overall yield of 18 per cent if rental growth is 7.65 per cent per annum. The argument seems to have gone full circle, since it has already been noted that the rational model has no device for doing this.

However, it is suggested that it might be possible, by utilising the formula for calculating rental growth, to discover the appropriate capitalisation rate.

$$g = \left[\left(\frac{d-y}{d} \right) \left((1+d)^t - 1 \right) + 1 \right]^{\frac{1}{5}} - 1$$

$$\text{or } (1+g)^5 = \left[\left(\frac{d-y}{d} \right) \left((1+d)^t - 1 \right) + 1 \right]$$

In this instance, g, t and d are known; y, the capitalisation rate, is unknown.

Thus

$$(1 + 0.0765)^5 = \left[\left(\frac{0.18 - y}{0.18}\right)\left((1.18)^5 - 1\right) + 1\right]$$

$$1.446 = \left[\left(\frac{0.18 - y}{0.18}\right)\left(2.288 - 1\right) + 1\right]$$

$$0.446 = \left(\frac{0.18 - y}{0.18}\right)(1.288)$$

$$0.346 = \frac{0.18 - y}{0.18}$$

$$0.0623 = 0.18 - y$$

$$y = 0.1177$$

Therefore a rack rented capitalisation rate of 11.77 per cent is required to produce an overall yield of 18 per cent, if rental growth is 7.65 per cent per annum.

The rational model in its original form now produces the following valuation:

Rent received	£45 000 pa			
less Rent paid	£14 000 pa			
Profit rent	£31 000 pa			
YP in perpetuity at 18 per cent	5.555		£172 205	
Less Rent received	£45 000 pa			
less Rent paid	£14 000 pa			
Profit rent	£31 000 pa			
YP in perpetuity at 18 per cent	5.555			
× PV of £1 in 3 years at				
18 per cent	0.609	3.383	£104 873	£ 67 332
Plus Rent received	£65 000 pa			
less Rent paid	£14 000 pa			
Profit rent	£51 000 pa			
× Amount of £1 in 3 years at				
7.65 per cent	1.2475			
Estimated profit rent	£63 622 pa			
In 3 years				
YP in perpetuity at 11.77 per cent	8.496			

× PV of £1 in 3 years at 18 per cent	0.609	5.174	£329 180	

Less Rent received	£65 000 pa	
less Rent paid	£14 000 pa	
Profit rent	£51 000 pa	
× Amount of £1 in 18 years at 7.65 per cent	3.769	

Estimated profit rent in 18 years		£192 219 pa		
YP in perpetuity at 11.77 per cent	8.496			
× PV of £1 in 18 years at 18 per cent	0.051	0.433	£ 83 231	£245 949
		Capital value		£313 281

say £315 000

This shows promise, but there still remains the problem of the differing growth prospects of head lease and sub-lease rents. The above valuation ignores this, but it may be taken into account by using rational format amended so that growth potentials are separately identified.

Rent received		£45 000 pa		
less Rent paid		£14 000 pa		
Profit rent		£31 000 pa		
YP in perpetuity at 18 per cent		5.555	£172 205	
Less Rent received		£45 000 pa		
less Rent paid		£14 000 pa		
Profit rent		£31 000 pa		
YP in perpetuity at 18 per cent	5.555			
× PV of £1 in 3 years at 18 per cent	0.609	3.383	£104 873	£ 67 332
Plus Rent received		£65 000 pa		
× Amount of £1 in 3 years at 7.65 per cent		1.2475		
Estimated rental value in 3 years		£81 087 pa		
YP in perpetuity at 11.77 per cent	8.496			
× PV of £1 in 3 years at 18 per cent	0.609	5.174	£419 544	
Less Rent paid		£14 000 pa		
YP in perpetuity at 18 per cent	5.555			

x PV of £1 in 3 years at 18 per cent	0.609	3.383	£ 47 362	£372 182
Less Rent received		£65 000 pa		
x Amount of £1 in 18 years at 7.65 per cent		3.769		
Estimated rental value in 18 years		£244 985 pa		
YP in perpetuity at 11.77 per cent 8.496				
x PV of £1 in 18 years at 18 per cent	0.051	0.433	£106 078	
Less Rent paid		£14 000 pa		
YP in perpetuity at 18 per cent 5.555				
x PV of £1 in 18 years at 18 per cent	0.051	0.283	£ 3 962 £102 116	£270 066

Capital value £337 398

say £340 000

Summary

There is no difficulty in reconciling the valuations of the freehold interest produced by the two models — both are approximately £740 000.

It is in the valuation of the leasehold interest that agreement is difficult to achieve.

The rational model has been the target of some criticism, but it appears to be the application of the model which may be at fault, rather than the reasoning behind it.

The basic problems are:

(i) Valuation of the profit rent, so that the differing growth potential of head lease and sub-lease rents are not taken into account.
(ii) Determination of the appropriate rack rented capitalisation rate necessary to produce the required risk adjusted op ortunity cost of capital at a particular rate of rental growth.

Once these difficulties have been overcome, the rational model produces valuations of the head leasehold interest at discount rates of 12 per cent and 18 per cent, which agree with valuations produced by a real value approach. Because they agree, is it reasonable to accept that the valuations are correct? Since both methods are claimed to be shortened discounted cash flow calculations, the validity of the valuations may be checked by providing discounted cash flow valuations of the head leasehold interest at 12 per cent and 18 per cent.

It is appreciated that, in an examination, lack of time would probably preclude such comparison, but it is included here to assist in an understanding of the principle.

Valuation of head leasehold interest using a discounted cash flow approach

(i) *Discount rate 12 per cent*

Year	Rental value £	Amount of £1 at 7.65 per cent	Inflated rental value £	Rent paid £	Profit rent £	YP at 12 per cent	PV at 12 per cent	Present value £
1-3	45 000	1	45 000	14 000	31 000	2.402	1	74 462
4-8	65 000	1.2475	81 087	14 000	67 087	3.605	0.712	172 196
9-13	81 087	1.446	117 252	14 000	103 252	3.605	0.404	150 378
14-18	117 252	1.446	169 546	14 000	155 546	3.605	0.229	128 410
					Capital value			£525 446

say £525 000

(ii) *Discount rate 18 per cent*

Year	Rental value £	Amount of £1 at 7.65 per cent	Inflated rental value £	Rent paid £	Profit rent £	YP at 18 per cent	PV at 18 per cent	Present value £
1-3	45 000	1	45 000	14 000	31 000	2.174	1	67 394
4-8	65 000	1.2475	81 087	14 000	67 087	3.127	0.609	127 757
9-13	81 087	1.446	117 252	14 000	103 252	3.127	0.266	85 883
14-18	117 252	1.446	169 546	14 000	155 546	3.127	0.116	56 422
					Capital value			£337 456

say £340 000

　　　The discounted cash flow calculations confirm the valuations of the head lease-hold interest produced by both real and rational models. It is apparent that these models, when correctly applied, are workable alternatives to a discounted cash flow approach, although great care must be taken when applying the rational model, particularly in the case of reversionary leaseholds. This does also tend to be a rather cumbersome and lengthy calculation.

APPENDIX: MODERN VALUATION TECHNIQUES

Derivation of 3 YP formula used in real value approach

This is derived from the summation of the present value of all the income flows, divided into 'blocks' of value according to the length of time between rent reviews.

Assumptions: Rent = £1 pa.

Interval between rent reviews = t years.

Total length of investment = n years.

Equated yield = e.

Inflation risk free yield = i.

(1) $\underset{\text{at } e}{S = \text{YP } t \text{ years}} \left(1 + \dfrac{1}{(1+i)^t} + \dfrac{1}{(1+i)^{2t}} + \dfrac{1}{(1+i)^{3t}} + \dfrac{1}{(1+i)^{4t}} + \ldots + \dfrac{1}{(1+i)^{n-t}}\right)$

For ease of calculation, but still retaining equality, both sides of the equation are multiplied by

$\dfrac{1}{(1+i)^t}$ giving:

(2) $S\left(\dfrac{1}{(1+i)^t}\right) = \underset{\text{at } e}{\text{YP } t \text{ years}} \left(\dfrac{1}{(1+i)^t} + \dfrac{1}{(1+i)^{2t}} + \dfrac{1}{(1+i)^{3t}} + \ldots + \dfrac{1}{(1+i)^n}\right)$

Equation (2) is subtracted from equation (1)

$$S - S\left(\dfrac{1}{(1+i)^t}\right) = \left(1 - \dfrac{1}{(1+i)^n}\right) \text{YP } t \text{ years at } e$$

$$S\left(1 - \dfrac{1}{(1+i)^t}\right) = \left(1 - \dfrac{1}{(1+i)^n}\right) \text{YP } t \text{ years at } e$$

$$S = \dfrac{1 - \dfrac{1}{(1+i)^n}}{1 - \dfrac{1}{(1+i)^t}} \times \text{YP } t \text{ years at } e$$

Numerator and denominator are now divided by i, giving:

$$S = \dfrac{\dfrac{1 - \dfrac{1}{(1+i)^n}}{i}}{\dfrac{1 - \dfrac{1}{(1+i)^t}}{i}} \times \text{YP } t \text{ years at } e$$

$$\dfrac{1 - \dfrac{1}{(1+i)^n}}{i} = \text{YP } n \text{ years at } i$$

$$\dfrac{1 - \dfrac{1}{(1+i)^t}}{i} = \text{YP } t \text{ years at } i$$

Thus

$$\frac{1 - \dfrac{1}{(1+i)^n}}{i} \times \text{YP } t \text{ years at } e$$
$$\overline{\dfrac{1 - \dfrac{1}{(1+i)^t}}{i}}$$

$$= \frac{\text{YP } n \text{ years at } i}{\text{YP } t \text{ years at } i} \times \text{YP } t \text{ years at } e$$

or, as expressed in the text:

$$\text{YP } n \text{ years at } i \times \frac{\text{YP } t \text{ years at } e}{\text{YP } t \text{ years at } i}$$

Equated yield

This may be defined as the overall yield from an investment, including rental and capital growth. It is used to value a fixed, or inflation prone income.

Inflation risk free yield

This is the yield used to value an income that can rise to meet any reduction in the value of that income caused by inflation.

It would be used to value a completely inflation-proof income.

10   Specialised Properties

This chapter deals with the valuation of properties of a specialised nature, namely:

(i) agricultural properties
(ii) sewers, pipelines and electricity easements and wayleaves
(iii) minerals
(iv) garages and petrol stations
(v) licensed premises.

In each section it has not been possible to deal with every valuation situation which may arise, and students are referred to the bibliography for further reading.

AGRICULTURAL PROPERTIES

The main legislation affecting the valuation of agricultural properties is the *Agricultural Holdings Act 1986*, which repeals much of the earlier legislation, including the *Agricultural Holdings Act 1948*.

Valuations may be required for sales with vacant possession and sales of tenanted farms, often to the sitting tenant. These are illustrated in Question **10.1**.

The vacant possession market has been slowed down by the small number of farms available compared with the early 1980s, but there is evidence of a recovery particularly in the dairy sector with milk quotas. These quotas may account for 50 per cent of the price when attached to bare land. Land of areas in the region of 250 hectares with farmhouses may continue to be in demand. Dairying and the production of sheep has a fairly prosperous future, but the pig industry may be risky due to the large increases in the costs of certain feeds. However, breeding techniques are becoming more sophisticated with the aid of computer control and there is at least one British farm with 3000 pigs in breeding at any particular time.

Cattle prices have reached unprecedented levels but it is doubtful whether these can be sustained.

The success of the cereal harvest is influenced by policies of the European Economic Community (as is all farming), one of the problems being that the European harvest may be in excess of Community requirements. This is exacerbated by the yield per hectare being doubled in the past ten years, because of the discovery of a new type of wheat and the increased use of fungicides. (The cost per hectare for fungicides and weedkillers may be in the region of £250 per annum.)

There may be moves to diversify into non-agricultural enterprises and there are now grants available towards feasibility studies into diversification plans.

The investment market in tenanted farms has improved slightly, with most sales being private with yields varying from 6 to 8 per cent. The sitting tenant may be the most likely purchaser, sale prices being dependent upon the respective bargaining strengths of the parties, but likely to be 30 to 50 per cent of the vacant possession value.

Students undertaking rural estate management courses and options will be involved in tenant right valuations, which is beyond the scope of this chapter (refer to *Agricultural Valuations* by R. G. Williams, published by Estates Gazette, 1985).

AGRICULTURAL PROPERTIES – QUESTION

10.1. A farm in the Midlands, 80 hectares, with a well-maintained farmhouse and buildings, is for sale. The land is served by mains water and is level and has good quality loam. The land is used for a mix of arable and dairy, with a substantial milk quota.

Value the farm for sale on the assumption that it is to be sold:

(i) with vacant possession;
(ii) as an investment, subject to an agricultural tenancy – the tenant pays a rent of £70 per hectare with a rent review in 2 years' time;
(iii) to the sitting tenant.

You may assume any necessary facts to produce an answer.

AGRICULTURAL PROPERTIES – SUGGESTED ANSWER

Question 10.1

(i)			
	80 hectares at £4 500 [see note 1]		£360 000
Less	Drainage rates say	£200 pa	
	YP in perpetuity at 5 per cent	20	£ 4 000
	Capital value		£356 000

Note

1: The value of £4 500 per hectare is based on average sale prices for the Midlands as at 1 October 1988.

(ii) Rent receivable

80 hectares at £70		£5 600 pa	
Less Drainage rates as before	£200		
Management say 15 per cent [see note 1]	£840	£1 040 pa	
		Net income £4 560 pa	
YP 2 years at 6 per cent [see note 2]		1.833	£ 8 358
Reversion to net rack rental value			
80 hectares at £100		£8 000 pa	
YP in perpetuity deferred 2 years at 6 per cent [see note 2]		14.833	£118 664
Capital value			£127 022
say £127 000			

Notes

1: An allowance of 15 per cent for land agents fees is considered reasonable.

2: A yield of 6 per cent is based upon transactions recorded in the Midlands in 1988.

(iii) The valuations prepared in parts (i) and (ii) show a capital value with vacant possession of £356 000 and a capital value subject to the agricultural tenancy of £127 000. The sitting tenant is a special purchaser who, on acquisition of the freehold, will benefit from marriage gain. The price to be paid would be dependent upon the respective bargaining powers of landlord and tenant and influenced by the landlord's need to sell. One might suggest a half-way point of say £241 000 but market evidence suggests that sitting tenant values are 30 to 50 per cent of vacant possession value — 40 per cent of vacant possession value would give a sale price of £142 400.

SEWERS, PIPELINES AND ELECTRICITY EASEMENTS AND WAYLEAVES

This section deals with the basis of compensation where an installation is laid in, on, over or under land, and will include water pipes, sewers, gas mains, electricity cables and overhead lines, oil pipelines and pipes carrying materials such as chalk and ethylene.

The term 'easement' is conveniently used to mean the width in which a pipe is laid, although this is not its strict legal meaning.

The situations usually found in examination questions involve:

(i) Sewers and water mains. Water authorities are empowered by the provisions of the *Public Health Act 1936* to lay public sewers and the *Water Act 1945* gives them powers to lay water mains. The *Water Act 1973* puts them under an obligation to provide and maintain a sewerage system. Water authorities are not required to negotiate easements; they need to give reasonable notice of their intention to lay a sewer and have a right of entry.

(ii) Gas pipes. The *Gas Act 1972* formed the British Gas Corporation, who had a duty to "develop and maintain an efficient, co-ordinated and economical system of gas supply." They do not have power of entry by right, but they can apply to the Secretary of State for compulsory purchase powers to acquire any right in land, including an easement.

(iii) Electricity mains and cables. Area electricity boards are responsible for distributing electricity to consumers within their area, and have a duty to lay suitable distribution mains. They do not have a right of entry, and have to obtain consent to enter and lay a pipe, by compulsory powers, if necessary.

(iv) Oil pipelines. The *Pipelines Act 1962* establishes a procedure whereby a company may apply to the Minister for compulsory purchase powers, after attempting to negotiate with landowners for the required rights. If this procedure fails, the company may then apply for a Compulsory Purchase Order or a Compulsory Rights Order. The Compulsory Rights Order gives the company the rights to lay a pipe, but does not amount to the purchase of an interest in land. If land is required for ancillary buildings, a Compulsory Purchase Order may be used.

In all cases, the payments made by acquiring authorities may be classified under two distinct headings:

(i) Consideration. This is for the acquisition of the easement, and there are no statutory guidelines as to the width of land to be allowed for payment, but narrow width easements should not be supported. A guideline may be a width which allows the Authority to enter upon their easement to carry out repairs without going outside the easement, and the common allowance is 10 feet either side of the pipe, say 22 feet (6.7 metres).

   The consideration is based on a proportion of freehold land value relative to the interest to be vested in the acquiring authority. Water authorities generally pay 50 per cent of freehold value plus allowances for manholes and other structures above and below ground. This follows guidelines laid down in *Markland and Felthouse v. Cannock RDC 1973*. Gas, electricity and oil undertakings seem more generous, allowing 75 per cent of capital value plus surface installations. British Gas negotiate a payment to be made for major pipelines with the Country Landowners' Association and the National Farmers' Union. At the time of writing, this is approximately £2 per metre run for a 22 feet easement. In the case of tenanted land, this would be split 67p to the tenant and £1.33 to the landowner.

The comparable allowances may be:

Electricity — £1.80 per metre run (split between tenant and landowner).
Oil — £7.20 per metre run (split between tenant and landowner). These rates, generous compared with other undertakers, are agreed for individual pipelines, allowing for increases in accordance with the Retail Prices Index.
Water — £1.40 per metre run. This is all paid to the landowner, the tenant's only financial claim being for temporary losses.

(ii) Compensation. The claim should be based on two principles:
(a) The claimant is left in no worse a position than he would have been had the event not taken place and where there is no other means of restitution then compensation by money shall be made, and
(b) it is the claimant's duty to mitigate wherever possible any claim that may arise.

The heads of claim for compensation which might be considered are shown in the answer to Question **10.2**.

In all cases, professional fees are paid by the acquiring authority.

SEWERS, PIPELINES AND ELECTRICITY EASEMENTS AND WAYLEAVES — QUESTIONS

10.2. (a) Calculate the value per metre run for a sewer easement in agricultural land worth £6 000 per hectare; the easement being 6 metres wide.

(b) Your client, who owns a large mixed farm, has been served with a Notice to Treat in respect of an underground electricity cable wayleave, and you have been instructed to prepare the claim for compensation.

Describe the items that might be included, discussing the main points to be taken into account.

10.3. Calculate the payment to be made to the owner of 2 hectares of land, which has planning permission for the erection of 40 detached houses. A sewer, length 300 metres, runs along the edge of the plot and a 10 m wide easement is sought by the local authority. It is possible to build all 40 houses on the site, but 5 plots will be adversely affected, their value being diminished by 10 per cent.

Show two methods of calculating the claim making all necessary assumptions.

SEWERS, PIPELINES AND ELECTRICITY EASEMENTS AND WAYLEAVES – SUGGESTED ANSWERS

Question 10.2

(a) Assuming that the acquiring authority allow 50 per cent of the value of £6 000 per hectare for the easement, then the value per metre run

$$= 50 \text{ per cent of £6 000} \times \frac{6}{10\,000} \quad [\text{see note 1}]$$

$$= \text{£1.8 per metre run}$$

Note

1: 6 m^2 is 0.0006 ha (10 000 m^2 = 1 ha).

(b) The payment for consideration for the easement, on the assumption that the authority would allow 75 per cent of the capital value of the land, would be similar to the above:

$$75 \text{ per cent of £6 000} \times \frac{6}{10\,000}$$

$$= \text{£2.7 per metre run}$$

If this is a low voltage cable, an annual wayleave agreement may be used, but for a high voltage cable, a formal easement would be required.

The compensation for losses may take account of the following:

(i) Loss of crops. This is calculated as the profit that might have been obtained from the area while cable laying operations are carried out. It may be based on the average crop over the farm, but care should be taken in case this particular land is of better quality than the average.

There may also be a claim for future loss of crops, dependent upon the standard of reinstatement.

Percentage allowances for loss over say 3 years may be taken, perhaps 50 per cent for the first year, 20 per cent for the second year and 10 per cent for the third year This is rather debatable, and would be subject to negotiation.

(ii) Loss of pasture. This, again, would be loss of profits on the working area.

(iii) Drainage. It is possible that, during the works, land drainage is disturbed and also the route of the cable may cut across land drains. The claimant should receive the cost of rectifying this situation.

(iv) Effect on soil. The structure of the soil may be damaged as a result of the cable laying, affecting its fertility. Any cost of upgrading the soil to its former state, such as the application of fertilizers, is claimable.

Surplus soil should not be removed as it may be required, when the soil covering the cable has settled, as a 'top-up'.

(v) Fencing and hedges. Fencing may be easily reinstated, but hedges will take time to mature and may require additional maintenance for several years.

(vi) Effect on livestock. The noise generated by the execution of the works may adversely affect livestock and reduce yields of eggs and milk. To substantiate a claim under this head, it would be necessary to have records available of livestock performance pre-cable laying and during cable laying.

(vii) Severance. If the works sever a field, and suitable alternative access is not available or less convenient, a larger area than the working area will be subject to a claim.

(viii) Injurious affection. This includes any item where the value of the remaining land is adversely affected by the cable.

(ix) Disturbance. This head might include dust and noise from the workings, interference with privacy and extra work incurred by the landowner due to the carrying out of the works.

When the claim is settled, it should not be closed entirely because some defects, particularly in land drainage, may not be apparent for several years.

Question 10.3

It is necessary to assume a capital value per hectare for residential building land. This may vary from £4.5 m per hectare in Inner London to £400 000 per hectare in the North West of England. For this answer, a typical value for the Midlands, £800 000 per hectare, is assumed.

If a 'before' and 'after' valuation is adopted, then the claim would be:

'Before' — 2 ha x £800 000 = £1 600 000

'After' — Value of 35 plots   £1 400 000

$$\text{Value of 5 plots} = \frac{£1\,600\,000}{40} \times 5 \times 0.9 \qquad £\ 180\,000$$

[see note 1]   £1 580 000

Diminution in value   £   20 000

Note

1: Each plot has a value of £40 000. 5 plots have 90 per cent of their original value, i.e.

$$£40\,000 \times 5 \times 90 \text{ per cent} = £180\,000$$

If the claim is calculated as a proportion of capital value, the calculation would be:

Length of pipeline	300 m x
Width of easement	10 m
Easement area	3 000 m²

This is 0.3 ha × £800 000 = £240 000. If 50 per cent of capital value is allowed —
this claim is 50 per cent, i.e. £120 000.

Clearly, the 'before' and 'after' valuation gives the smaller figure, although this
may not always be so.

MINERALS

Mineral valuation is a specialised field requiring detailed knowledge of minerals and
their methods of extraction and markets.

The most specific definition of 'minerals' may be found in *Section 29 and
Schedule 6 of the Finance Act 1970*: "All minerals and substances in or under land
which are ordinarily worked or removed by underground or surface working but
excluding water, peat, topsoil and vegetation."

Minerals may be classified as follows:

(i) Those used in the construction industry. There is evidence that quarries are
increasing production levels because of major road schemes and a buoyant housing
market. This is reflected in an increase in royalties currently at 40p per m^3 for
common clay and higher for fireclay, gravel, sand and limestones.

(ii) Those connected with the energy industry. The operators of private opencast
coal sites and small mines require licence fees from British Coal. These are currently
charged at about £12 per tonne for opencast workings and 75p per tonne for small
mines. Payments made by private operators for surface rights overlying coal licensed
areas are increasing.

(iii) Agricultural minerals. The market for chalk is fairly steady, the highest values
being in the South East of England.

(iv) Chemical industry minerals. The ball clay and china clay industries are healthy
in the South West of England, but steady elsewhere.

Questions on minerals may require an explanation of royalties, dead or certain
rents and short-workings clauses, and the calculation of royalties and freehold valuations as required in Question **10.4.**

MINERALS — QUESTION

10.4. (a) Explain how income from mineral-bearing land may be structured in a
mining lease.

(b) Calculate the royalty payment to be paid to the freehold owner by a
mineral operator. The estimated output of the quarry is 30 000 tonnes per
annum and the market price of chalk is £2.50 per tonne. The operator will
invest £125 000 in the operation, and his working costs will be £30 000 per
annum.

(c) Value the freehold interest in an 8 hectare quarry let for 40 years or until the quarry is exhausted. The total chalk content is considered to be 100 000 tonnes per hectare and the anticipated rate of working is 25 000 tonnes per annum.

The lease requires a surface rent of £100 per hectare and a certain rent of £2 500 per annum. The royalty payment has been agreed at 30p per tonne.

MINERALS – SUGGESTED ANSWER

Question 10.4

(a) Mineral leases usually provide for three kinds of rent:

(i) Surface rent, which is a payment in respect of the surface occupied.
(ii) Royalty payments, which are based on the output of mineral worked, usually expressed as so much per tonne.
(iii) A minimum rent per annum to be paid irrespective of the amount of mineral extracted (sometimes called a 'certain' or 'dead' rent).

If rent is based entirely on output, the owner's cash flow depends on the operator's ability to extract each year. The operator may initially hold the land to ensure future continuity of supply, so he will be obliged to pay the minimum rent whether or not minerals are actually worked. This rent is normally offset against royalties payable in each year. Some leases also incorporate a 'short-workings' clause, in an attempt to recompense the operator where he extracts material in a year, which has a royalty value less than minimum rent. The deficit is offset against royalty payments in excess of the minimum rent in a subsequent year.

Example

The minimum rent is £5 000 per annum. During the first year of a mining lease, there is no extraction of minerals. However, the operator pays £5 000.

During the second year, minerals worth £12 000 in royalties are extracted. There is, however, a 'short-workings' clause in the lease, so that the rent paid is £12 000 – £5 000 = £7 000.

(b) Selling price per tonne =   £2.50

Less

(i) Working costs per tonne [see note 1]

$$\frac{£30\,000}{30\,000} =$$ £1.00

(ii) Interest on capital [see note 2]

$$= \frac{£125\,000 \times 12 \text{ per cent}}{30\,000} =$$ £0.50

                                                                       £1.50

                                   Residue   £1.00
            Operator's profit say 50 per cent [see note 3]   £0.50

                              Royalty per tonne   £0.50

Notes

1: The first deduction to be made from the selling price per tonne is for working expenses expressed as an amount per tonne.

2: The capital invested in the project might have earned 12 per cent if invested elsewhere.

3: The residue has been divided equally between the operator and owner in this example, but this would not always be the case and would be subject to negotiation.

(c) *Life of reserves:*

Total reserves 8 hectares × 100 000 tonnes = 800 000 tonnes

$$\text{Annual output} = \frac{800\,000}{25\,000} = 32 \text{ years}$$

Annual royalty payment   = 25 000 tonnes × 30p

                              = £7 500

Less Certain rent   £2 500

       Net payment   £5 000 per annum

Valuation

(i) Surface rent 8 hectares × £100 =	£ 800	
YP 32 years at 15 per cent		
[see note 1]	6.59	£ 5 272
(ii) Certain rent	£ 2 500	
YP 32 years at 15 per cent and		
3 per cent (tax 40 per cent)		
[see note 2]	5.5	£13 750
(iii) Net payment for royalties	£ 5 000	
YP 32 years at 20 per cent and		
3 per cent (tax 40 per cent)		
[see note 3]	4.315	£21 575
Total value		£40 597
say	£40 600	

Notes

1: The surface rent has been capitalised with single rate Years' Purchase to reflect that there may be some residual value in the land. There might ultimately be potential for tipping purposes or leisure uses. It may be possible to include in the valuation the future capital value of the site deferred 32 years, with an allowance for reinstatement costs.

2: The certain rent of £2 500 is guaranteed income for 32 years, so that 15 per cent yield is appropriate.

3: The net payment for royalties may be regarded as a riskier 'top slice' of income, so that the yield is increased to 20 per cent.

GARAGES AND PETROL STATIONS

The types of garages and petrol stations may fall into a number of different categories:

(i) Those which sell petrol, but have no garage repair or showroom facilities. These vary from small town service stations to motorway areas, and additional services such as car wash facilities, shops and restaurants will depend upon location and trade potential. Oil companies are looking for such properties more as a medium for obtaining outlets for their fuels than for investment; this gives rise to profits from refining and wholesaling whereas a dealer will profit from retailing. They probably require annual throughputs of 4 to 4.5 million litres of fuel to induce them to purchase or lease petrol stations. They may prefer purchasing to leasing. In the latter case, an oil company may pay a 'free rent' which includes 'tied rent' from a dealer and an overbid based on the company's margins. The dealer will obviously pay his rent out of retailing profits.

(ii) Those with petrol sale and car repair facilities (and perhaps car sales).

These may be in private ownership with an annual throughput of fuel less than 2.25 million litres.

(iii) Garages with repair facilities only, which will inevitably be in private ownership. There may also be provision for vehicle sales.

(iv) Motorway service areas. These may have an annual throughput of 20 million litres.

The valuation of petrol stations may be assisted by good comparables, but the practical difficulty often encountered is the variable nature of layout, accessibility, petrol pricing and dealer operations even with garages in close proximity to each other. The re-emergence of trading stamps and the Government's campaign to encourage car-owners to switch to lead-free petrol may be influencing factors.

It is likely that a valuer may need to estimate throughput when preparing a capital valuation of a petrol station, dependent on traffic flows, density of population and distance between service stations. There will be a need also to distinguish between 'market rents' and 'tied rents' (mentioned earlier). Open market rents (also referred to as 'free rents') may be expressed as a rate per litre related to the cost of producing petrol and other lubricants and marketing profits. This is variable according to throughput and geographical location. In the South, 1.5p per litre may be typical and in the Midlands and North 1p per litre.

The 'tied rent' is based on the dealer's margin and may be in the region of 0.5p per litre. The dealer's margin is worked from the addition of retail pump prices, and rebates with deductions for 15 per cent Value Added Tax and the wholesale price.

Questions **10.5** and **10.6** illustrate many of the issues discussed above.

GARAGES AND PETROL STATIONS – QUESTIONS

10.5. Value a garage and petrol station in a Nottinghamshire town with anticipated petrol sales of 3 million litres per annum. The garage has a small shop on the forecourt and car repair facilities employing 4 mechanics.

Assume any other information necessary to produce an answer.

10.6. Value a petrol station on a trunk road situated in the South East with anticipated petrol sales of 6.5 million litres per annum.

Assume any other information necessary to produce an answer.

GARAGES AND PETROL STATIONS – SUGGESTED ANSWERS

Question 10.5

It is necessary to calculate the rental value per annum and then capitalise this to arrive at capital value. Different yields may be adopted for the forecourt and the remainder. In the following calculation typical areas and rents per m^2 have been assumed for the buildings.

Forecourt –	3 m litres at 1p [see note 1]	=		£30 000 pa
Shop	40 m^2 x £30	= £1 200		
Workshop	200 m^2 x £30	= £6 000		
Stores	50 m^2 x £12	= £ 600		
Lubrication bays	75 m^2 x £30	= £2 250		
Parking	350 m^2 x £ 4	= £1 400	£11 450 pa	
	Rack rental value		£41 450 pa	

Capital valuation

Forecourt – rack rental value	£30 000 pa	
YP in perpetuity at 9 per cent [see note 2]	11.11	
		£333 300
Remainder – rack rental value	£11 450 pa	
YP in perpetuity at 12 per cent [see note 2]	8.334	£ 95 425
Capital value		£428 725

Notes
1: It is considered that 1p per litre based on petrol sales would be an appropriate rent for a Midlands town.
2: The yields for capitalisation vary from 9 per cent for the forecourt and 12 per cent for the remainder to reflect the fact that the attractiveness of the investment is related to the potential growth in petrol sales.

A 'tied rent' may be based on say 0.3p per litre, i.e. 3 m litres x 0.3p = £9 000 per annum plus £11 450 as before = £20 450 per annum.

Question 10.6

Forecourt – 6.5 m litres at 1.5p [see note 1] =	£	97 500 pa
Shop 75 m^2 x £50 =	£	3 750 pa
Rack rental value [see note 2]	£	101 250 pa
YP in perpetuity at 8 per cent		12.5
Capital value		£1 265 625

Notes
1: It is considered that 1.5p per litre based on petrol sales is a reasonable rent for this location.

2: The shop rental is capitalised at the same yield as the forecourt rental, as its sales may be greatly influenced by petrol sales and it is regarded as an integral part of the forecourt.

A 'tied rent' may be based on say 0.5p per litre, i.e. 6.5 m litres × 0.5p = £32 500 per annum plus £3 750 as before = £36 250 per annum.

LICENSED PREMISES

Licensed premises may be split between hotels and public houses. Hotels are valued on the basis of open market value, recognising that they are specialised buildings with considerable trading potential. Their valuation may be based on a profits method, and an example is provided in the answer to Question 5.7, which considers the rating of a licensed hotel. The methods of valuing public houses must take into account the level of trading activity, as it does not form the basis of a separate goodwill calculation but is reflected in the capital value. Brewers may only be interested in purchasing public houses with potential growth in trade and there is evidence that they are selling off their smaller houses. Some independent brewers, however, are attempting to build up a chain of public houses, particularly those retailing real ale.

When considering rent in valuations, it must be established whether or not the tenant is paying a 'tied rent', which might be regarded as being less than an open market rent, because the tenant is tied to selling the brewer's products.

The methods available include a brewer's profit basis as indicated in Commissioners of Inland Revenue v. Allied Breweries (UK) Ltd (1982). The valuation in this case was based on the Kennedy Method established in *Ashby's Cobham Brewery Co. Ltd re The Crown, Cobham and Ashby's Staines Brewery Co. Ltd re The Hand and Spear, Woking (1906)*.

It is likely that a valuer will use an accounts method to value public houses as illustrated in Question 10.7.

If an open market rent has just been established, an investment method might be adopted.

LICENSED PREMISES – QUESTION

10.7. Construct an example to illustrate the capital valuation of a free public house, using appropriate figures.

LICENSED PREMISES – ANSWER

Question 10.7

It is assumed that the valuer would have access to trading figures and would establish receipts from sales of beer, spirits, cigarettes and refreshments, say £300 000 per annum. He is also able to establish from the accounts a net profit of £75 000 per annum.

A typical valuation would be:

Net profit		£ 75 000 pa
Plus Interest on loans [see note 1] say		£ 5 000 pa
		£ 80 000 pa
Less Interest on capital: [see note 2]		
Furniture, fixtures and fittings [see note 3] say	£ 7 500	
Stock in hand say	£10 000	
Day-to-day cash [see note 4] say	£15 000	
	£32 500	
Interest at 12 per cent		£ 3 900 pa
Adjusted net profit		£ 76 100 pa
Tenant's share say 50 per cent [see note 5]		£ 38 050 pa
Amount available for rent		£ 38 050 pa
YP in perpetuity at 7 per cent		14.28
Capital value say		£543 350

Notes

1: These loans are personal to the existing tenant and a prospective purchaser may be able to finance himself without loans.
2: Interest on capital tied up in the business is deducted to arrive at a profit level truly reflective of the business activity.
3: The estimation of the value of furniture, fixtures and fittings should allow for depreciation.
4: This is the cash necessary to keep the business going, and may be based on purchase of stock for a few weeks.
5: As in all profits methods, a percentage of adjusted net profit is deducted to arrive at the amount a tenant would be prepared to pay in rent. A 50:50 split is shown, but this could be 60:40 according to tenant's circumstances and the potential of the public house.

Bibliography

Baum, A., *Statutory Valuations*, Routledge and Kegan Paul (1983).

Booth, R. D., *Estate Manager's Rating and Tax Guide, Part I*, University of Reading (1984).

British Gas PLC, *Annual Report and Accounts* (1988).

Butler, D., *Applied Valuation*, Macmillan (1987).

Crosby, N., 'The Application of Equated Yield and Real Value Approaches to Market Valuation: 2', *Journal of Valuation*, Vol. 4 (1986).

Crosby, N., *A Critical Examination of the Rational Model*, University of Reading (1987).

Darlow, C. (Ed.), *Valuation and Investment Appraisal*, Estates Gazette (1983).

Davidson, A. W., *Parry's Valuation and Conversion Tables (10th Edition)*, Estates Gazette (1978).

Inland Revenue, *Property Market Report Number 49*, Surveyors Publications (1988).

ISVA, *A Guide to Asset Valuations (2nd Edition)*, ISVA (1988).

Kessler, J., 'Forewarned, Forearmed', *Taxation* (19 January 1989).

Lumby, S., *Investment Appraisal (3rd Edition)*, Van Nostrand Reinhold (UK) (1987).

Mellows, A. R., *Taxation of Land Transactions*, Butterworth (1982).

Powell, M. C., *An Examination of Some Inconsistencies Associated with Pipeline Compensation Payments*, Nottingham Polytechnic Dissertation (1988).

Rayner, M., *National and Local Taxation*, Macmillan (1978).

Rayner, M., *Asset Valuation*, Macmillan (1988).

Rees, W. H. (Ed.), *Valuation: Principles into Practice*, Estates Gazette (1988).

Richmond, D., *Introduction to Valuation*, Macmillan (1985).

RICS, *Guidance Notes on the Valuation of Assets, GN5, (2nd Edition)* (revised January 1989).

Sedgwick, J. R. E. and Westbrook, R. W., *The Valuation and Development of Petrol Filling Stations*, Estates Gazette (1969).

Soares, C., 'VAT on Property', *Taxation* (23 March 1989).

Trott, A. (Ed.), *Property Valuation Methods*, RICS Research Report (1986).

Westbrook, R. W., *The Valuation of Licensed Premises*, Estates Gazette (1983).

Westwick, C. A., *Property Valuation and Accounts*, ICAEW (1980).

Williams, R. G., *Agricultural Valuations — A Practical Guide*, Estates Gazette (1985).

Index

295